T0223818

Global Warming and the Future of the Earth

Global Warming and the Future of the Earth
Robert G. Watts

ISBN: 978-3-031-79416-2   paperback
ISBN: 978-3-031-79417-9   ebook

DOI: 10.1007/978-3-031-79417-9

A Publication in the Springer series

SYNTHESIS LECTURES ON ENERGY AND THE ENVIRONMENT: TECHNOLOGY, SCIENCE, AND SOCIETY # 1

Lecture #1

Series Editor: Frank Kreith, professor emeritus, University of Colorado

Global Warming and the Future of the Earth

Robert G. Watts
Tulane University

SYNTHESIS LECTURES ON ENERGY AND THE ENVIRONMENT: TECHNOLOGY, SCIENCE, AND SOCIETY # 1

ABSTRACT

The globally averaged surface temperature of the Earth has increased during the past century by about 0.7°C. Most of the increase can be attributed to the greenhouse effect, the increase in the atmospheric concentration of carbon dioxide that is emitted when fossil fuels are burned to produce energy. The book begins with the important distinction between weather and climate, followed by data showing how carbon dioxide has increased and the incontrovertible evidence that it is caused by burning fossil fuels (i.e., coal, oil, and natural gas). I also address the inevitable skepticism that global warming arouses and offer a number of responses to the global warming skeptics. After dealing with the skeptics, I analyze both the current and future effects of global warming. These future effects are based on scenarios or "storylines" put forth by the International Institute for Applied Systems Analysis. In closing, I address the controversial (and grim) suggestion that we have already passed the "tipping point," which is the time after which, regardless of our future actions, global warming will cause considerable hardship on human society. I intend this book to be approachable for all concerned citizens, but most especially students of the sciences and engineering who will soon be in a position to make a difference in the areas of energy and the environment. I have tried to frame the debate in terms of what the engineering community must do to help combat global warming. We have no choice but to think in terms of global environmental constraints as we design new power plants, factories, automobiles, buildings, and homes. The best thing for scientists to do is to present what we know, clearly separating what is known from what is suspected, in a non-apocalyptic manner. If matters are clearly and passionately presented to the public, we must be prepared to accept the will of the people. This presents the scientific community with an enormous responsibility, perhaps unlike any we have had in the past.

KEYWORDS

carbon dioxide, global warming, greenhouse effect, fossil fuels, energy, skeptics, impacts

Contents

Introduction

On June 23, 1988, Doctor James Hansen testified before a congressional committee that he believed with a "high degree of confidence" that the greenhouse effect had already caused global warming. After that testimony, there has been an increasingly acrimonious debate between those who see the problem as the most serious one facing humans today and those who refuse to believe there is any problem at all. Accusations and counteraccusations have spilled over into such august publications as *Nature* and *Science*. Some accused one of the principal authors of the Intergovernmental Panel on Climate Change (IPCC) Third Assessment Report[1] of changing the intent of that report to reflect much more confidence that warming has already been detected than many of the participating scientists are comfortable with. But the Fourth Assessment Report goes even further, stating that humans are responsible for global warming due to the emission of greenhouse gases (mostly carbon dioxide [CO_2]) with 95% confidence. On the other side of the debate, the doubters are often accused of being financially dependent (for research money) on the coal or oil industry. Ross Gelbspan, in his recent book *The Heat is On,*[2] implies that just about everyone who doubts the seriousness of global warming is being paid by the coal industry to obfuscate things. Environmental groups are accused of being alarmists, whereas scientists on the other side of the debate are accused of being only worried about their own self-interests rather than about future generations. Newspapers, television reporters, and newsmagazines love it when this happens. It makes for great stories. Kevin Sweeney issued a commentary (http://www.salon.com/news/feature/2001/03/29/kyoto/) calling President Bush's decision to pull out of the Kyoto Protocol "a national disgrace." Doctor Fred Seitz, president emeritus of Rockefeller University, states flatly that "Research data on climate change do not show that human use of hydrocarbons is harmful. To the contrary, there is good evidence that increased atmospheric carbon dioxide is helpful" (http://www.sepp.org/pressrel/petition.html). The average citizen is simply confused.

This is generally not a good way to inform the public about what scientists know about potentially important scientific questions. When scientific matters and science itself enter the political stage, particularly when scientists know (or hope) that their views will influence policy in important ways, there is a strong and perhaps natural urge for them to become ideologues and to emphasize that part of the science that supports their political views. Although scientists have an obligation to explain important discoveries to the public in ways that they can understand, telling exaggerated versions of the dangers of environmental problems may not convince people of the need to radically reconstruct government or change their behavior.

Several years ago, I was asked by a local environmentalist group to participate in a news conference heralding the dangers of greenhouse warming. During my conversation with the organizer, I was asked whether I was alarmed about the prospect of global warming. I replied that I was concerned, but not alarmed. I was quickly uninvited. It reminded me of a time when I was a postdoctoral fellow at Harvard and was asked to become a member of the Union of Concerned Scientists. I declined, saying that I did not like to be associated with groups that made blanket proclamations about things that I did not necessarily believe. Groups are like that. But it occurred to me when I spoke to the organizer of the news conference that it was no longer politically correct to be concerned. One must now be alarmed!

For the public to responsibly put a value on environmental concerns, it must be educated about the prospects of environmental degradation due to energy production, including possible climate change. There is, however, a danger that must be recognized at the outset. Education is not the same as indoctrination. In his book *Extraordinary Popular Delusions and the Madness of Crowds*,[3] Charles MacKay recalls Schiller's dictum: "Anyone taken as an individual is tolerably sensible and reasonable—as a member of a crowd, he at once becomes a blockhead." We need to avoid "crowd thinking" when we seek solutions to problems such as global warming. There are few guidelines as to how to do this. The average person, even those who think global warming is a problem, thinks of it as a long-term problem. Faced with the more immediate and visible problems of unemployment, poverty, famine, and war, the public tends to quickly tire of hearing about what they perceive as longer term, less certain, and certainly less visible problems such as global warming. Furthermore, it does not help to single out events such as a given very hot summer or a season of unusual floods and lay the blame definitely on global warming. Climate is noisy; it varies from year to year and from decade to decade. It is not unlikely for a hot summer to be followed by a couple of cool ones, and when that happens those who doubt that global warming is real will have a heyday. On the other hand, confronting problems such as the prospect of global warming can only be effectively done in a democratic society if the constituency (the public) is willing to confront the problem and endure the personal sacrifice that may be necessary to overcome it. The best thing for scientists to do is to present what we know, clearly separating what is known from what is suspected, in a nonapocalyptic manner. "Crowd thinking" tends to have a short lifetime.

If matters are clearly and impassionately presented to the public, we must be prepared to accept the will of the people. This presents the scientific community with an enormous responsibility, perhaps unlike any we have had in the past. This is particularly true of the engineering community, which has for the most part based designs almost entirely on such constraints as economy of sales, the immediate safety of the consumer (to prevent lawsuits, for example), and federal guidelines when they exist (and they almost always do). If global warming is a real threat, we now need to begin to think in terms of global environmental constraints as we design new power plants, factories, automobiles, buildings, and homes.

We need first to talk about weather and climate. Climate is not the same thing as weather, but climate affects weather. How does the climate machine work and how does it affect weather? What do we know about it and what are the limits of our knowledge? But there is more to the story. The environment, locally and globally, can be—is being—affected by the actions of people; few would argue against this. What are we doing that is likely to affect the climate? Is it necessarily bad? Can our industrial infrastructure, as well as our personal behavior, be changed in such a way that they are less environmentally destructive? If so, how, and at what cost? There is a need for the public to understand the whole of the current debate about climate change and its implications for our future.

One problem is that the science of climatic change is reported in specialist scientific journals (as it should be) in words, equations, and graphs that are largely impenetrable to the nonscientist, and often, even to scientists working in related fields. When extreme and, often, conflicting views are reported by the press or in popular magazines or books, the public tends to be dazzled and confused. What they tend to believe is that if such lettered experts disagree so widely, then they must all be either confused or making things up. There is a growing public perception that those who believe that the prospect of global warming is one of the great threats to future generations are "radical environmentalists," whereas those who do not believe the threat is real or serious are in bed with the coal companies.

Politicians are not helpful. Some recent statements by politicians regarding the Arctic National Wildlife Reserve (ANWR) serve to illustrate how public attitudes can be polarized by taking numbers out of context without giving the public supporting data. Awhile back, both Senator Tom Daschle and former Vice President Al Gore stated in speeches that the ANWR holds only enough oil to last the United States for 6 months. On the other hand, then-Senator John Breaux put the number much higher, perhaps 25 years. How can such diverse claims be justified? Actually, it is not very difficult if you do not tell where the numbers are coming from. A 1990s estimate of the amount of oil in the ANWR is 3.2 billion barrels. More recent estimates are between 5.7 and 16 billion barrels using currently available technology, and much more if drilling technology improves as expected. Total U.S. use of oil in 1999 was 20 million barrels per day. Imports from Saudi Arabia amounted to 1.566 million barrels per day. If you divide the smallest estimate of the available oil (3.2 billion barrels) by the largest estimate of oil use rate (20 million barrels per day) you get about 160 days, in line with the Daschle-Gore claims. On the other hand, if you divide the largest number (16 billion barrels) by the Saudi Arabia imports number (1.566 million barrels per day) you get about 28 years, in line with the Breaux claim. Daschle and Gore are against drilling in the ANWR, so they use the former figure. Breaux is in favor of drilling, so he uses the latter figure. But are they in favor of or against drilling because they believe the figures, or are they for or against drilling for other reasons and simply using the figures as smoke and mirrors? The American people deserve better.

Let us now look at the evidence of whether global warming is real, whether it has already occurred, and whether, if it is going to happen, it will be bad or good or benign. In this book, I will examine, in what I hope will be clear arguments understandable to the layman, the evidence for and against the idea that global warming due to the emission of CO_2 and other greenhouse gases into the atmosphere is a threat to the future of the planet. My aim is to present the evidence in a manner that allows people to make their own decisions about the threat and decide what if any difficult decisions we need to make as a society in the future. I will also give my own views and suggestions.

In addition, I have provided many references so that if the reader desires, he or she may go directly to the original source. In the words of Damon Runyon "You could look it up."

It will become clear to the reader that I believe that the problem of global warming is real and very serious. If it is, what solutions are available, short of shutting down the industrial infrastructure of the world? Energy use is necessary to run a prosperous and civilized society. Returning to a pre–industrial revolution lifestyle is simply not an option. One only needs to think of the difference in living conditions between developed and developing nations to see that to feed and house in reasonable comfort some 10 to 12 billions of people, one needs modern agricultural practice and a means for transporting food, as well as a reasonable manufacturing infrastructure at a minimum.

For many years I have time, I have been telling people during my many talks to professionals and lay audiences that the so-called tipping point, when our reluctance to stem global warming would surely lead to some very serious consequences regardless of our future actions, would come in 15 to 20 years. I now believe that we passed that point some years ago. I am alarmed! The late Doctor Ralph Rotty, one of my most important mentors, told me early in my career that I must not state the case of global warming so strongly that it turns people off. If you sound too apocalyptic, people will stop listening. Recent scientific evidence, however, has convinced me that the problem is so serious that scientists must sound the alarm loud and clear, and it must come from us, scientists who have seriously studied the subject. You will see why I feel this way when you read Chapter 3 (about observations of climate change) and Chapter 5 (about the consequences of future climate change). Much more observational data have come out in the scientific literature recently, and some are positively scary. In addition, all of the infrared herrings put forward by the so-called global warming skeptics are rather easily refuted, as I do in some detail in Chapter 4. The probability of dangerous sea level rise has increased substantially, as has the prospect of harm to the living environment, including sea creatures. There is an old Chinese curse "May you live in interesting times." I fear that my grandchildren, and perhaps even my children, will indeed live in interesting and environmentally disastrous times.

A wise man once said that if you think education is expensive, try ignorance.

If you think doing something about global warming will be expensive, try doing nothing. I fear that we are going to find out.

NOTES AND REFERENCES

1. "Intergovernmental Panel on Climate Change: The Third Assessment," in *Climate Change 2001: The Science of Climate Change*, Houghton, J.T., Meira Filho, L.G., Callendar, B.A., Harris, N., Kattemberg, A., and Maskell, K., Eds., Cambridge University Press, 1996.
2. Gelbspan, R., *The Heat is On*, Addison-Wesley, 1997.
3. MacKay, C., *Extraordinary Popular Delusions and the Madness of Crowds*, Farrar, Strauss & Giroux, 1932.

CHAPTER 1

Weather and Climate (and a Little History)

Weather is what you are now seeing when you look out of your window. It may be fair, cloudy, raining, or snowing, and it may be warm, mild, or cold. As I will explain later, it is not the same as climate.

1.1 WHY THE ATMOSPHERE FLOWS

We live at the bottom of a layer of gas that covers the Earth: the atmosphere. The Earth orbits the Sun about every 365 days and spins on an axis that is tilted about 23 degrees from the axis of its elliptic orbit (Figure 1.2). It is this tilt that is responsible for the change of seasons. When the northern hemisphere is tilted towards the Sun, it is northern hemisphere summer, and likewise for the southern hemisphere (Figure 1.2).

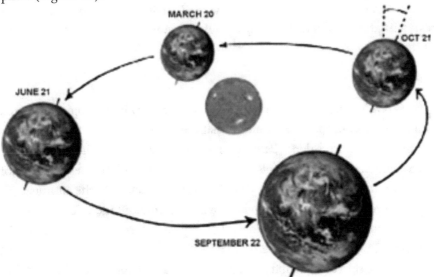

FIGURE 1.1: The Sun–Earth system. The Earth is spinning on its axis (in the counterclockwise sense when looking down from the north pole), which is tilted about 23 degrees. It also rotates about the Sun, making one complete orbit about every 365 days. The tilt of the Earth is responsible for the change of seasons as explained in the text.

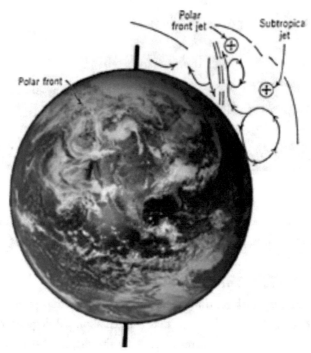

FIGURE 1.2: A schematic diagram of atmospheric motions. As the air in the Hadley cell moves toward the equator, the spin of the Earth about its axis moves the surface toward the right in the figure. An observer on the surface then feels a component of the wind coming from the east. Similarly, in the Ferrel cell in midlatitudes, the air tends to flow away from the equator and as the Earth's surface spins toward the right, there is a wind component from the west.

The Sun's rays impinge on the atmosphere nearly vertically in the tropics but at an increasing angle of incidence nearer the poles. This means that the regions closest to the equator receive more sunlight in a given area than regions closer to the poles, and as a result, they are warmer. Warm air is lighter than cooler air. It therefore tends to rise. As the warm, moist air near the equator rises, the water vapor in the air condenses into fine droplets, forming clouds. As the air nearest the equator rises, nearby air north and south of the equator must rush in to fill the void left behind. Thus, air from both sides of the equator converges toward the equatorial region, and meteorologists call this region the intertropical convergence zone. The rising air then turns poleward and descends, flowing downward toward the surface at around 30 degrees from the equator. This cellular motion is known as the Hadley cell and is illustrated in Figure 1.2. Because the descending air tends to be dried out, many of the Earth's deserts are located at latitudes near 30 degrees north or south of the equator.

Because the Earth is spinning on its axis, the air at the surface within the Hadley cell veers to the right toward the west. In other words, the Earth is spinning out from under the air as it turns on its axis, so that someone on the (moving) Earth experiences a wind from the east. The winds blowing toward the west are known as the trade winds. Poleward of the Hadley cell is the Ferrel cell, also illustrated in Figure 1.2. Air drawn down by the Hadley cell flows generally poleward near the surface, but because the Earth is rotating out from under these winds, we experience them veering toward the east. These midlatitude winds are known as the westerlies (from the west). Still further poleward, there is another weak cell: the polar cell. Again, the surface motion is generally toward the equator, but it veers to the right, so that the surface winds have an eastern component.

Near the equator, the surface winds are generally light, and for this reason, the region was named the doldrums (which means something like "low in spirit"). Similarly, in the region between the trades and the westerlies, the surface winds are light. In this region, in the days of sailing ships, vessels frequently became calmed for long periods, and it was named the horse latitudes (possibly because horses had to be eaten or thrown overboard when food and water shortages developed).

1.2   WHY THE OCEAN FLOWS

In part, the ocean surface is driven by the winds.[2] For example, look at a picture of the flow pattern in the north Atlantic Ocean (Figure 1.3). The trade winds blow the ocean surface water toward the west at low latitudes, whereas the westerlies blow the water toward the east at midlatitudes. At the same time, the rotation of the Earth forces the flow to "pile up" along the western boundary of the ocean (the eastern boundary of the continent), producing the Gulf Stream. Easterly (toward the west) winds at higher latitudes blow water to the west and down the east coast in the form of the Labrador Current. Similar wind-driven circulation patterns occur in other ocean basins. A very simplified pattern is shown in Figure 1.3. Note that the pattern is similar in the two hemispheres. The atmospheric winds generally blow the surface ocean currents toward the west near the equator and toward the east at midlatitudes. On the western sides of oceans (the eastern sides of continents), there are currents similar to the Gulf Stream: the Brazil Current off the east coast of South America, the Kuroshio Current off the east coast of Asia, the Mozambique Current off the east coast of Africa. There are large regions of slowly clockwise rotating water masses (called gyres) in the northern Atlantic and Pacific and large counterclockwise rotating water masses in the southern Atlantic and Pacific and in the Indian Ocean. In the southern hemisphere, where there is a clear ocean path that encircles the globe (no continental boundaries to stop the flow), the southern hemisphere westerlies create the Antarctic Circumpolar Current which, unimpeded by land masses, flows all the way around the globe, and the easterly winds below the southern hemisphere polar cell create the East Wind Drift near Antarctica, which also flows around the globe.

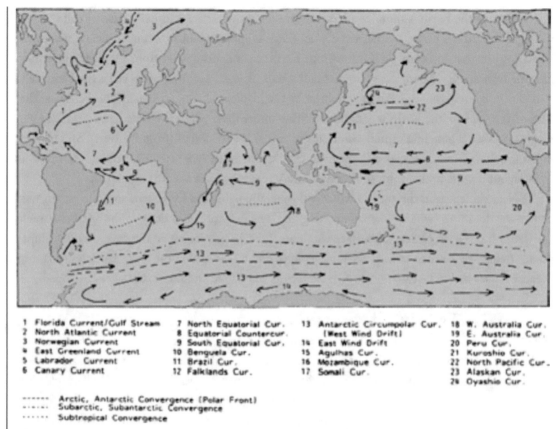

1 Florida Current/Gulf Stream	7 North Equatorial Cur.	13 Antarctic Circumpolar Cur.	18 W. Australia Cur.
2 North Atlantic Current	8 Equatorial Countercur.	(West Wind Drift)	19 E. Australia Cur.
3 Norwegian Current	9 South Equatorial Cur.	14 East Wind Drift	20 Peru Cur.
4 East Greenland Current	10 Benguela Cur.	15 Agulhas Cur.	21 Kuroshio Cur.
5 Labrador Current	11 Brazil Cur.	16 Mozambique Cur.	22 North Pacific Cur.
6 Canary Current	12 Falklands Cur.	17 Somali Cur.	23 Alaskan Cur.
			24 Oyashio Cur.

```
------  Arctic, Antarctic Convergence (Polar Front)
-.-.-.  Subarctic, Subantarctic Convergence
......  Subtropical Convergence
```

FIGURE 1.3: Flows in the ocean surface layer. The wind blows the surface water of the oceans toward the east under the Hadley cell and toward the west under the Ferrel cell. This leads to large rotating regions in the open ocean both north and south of the equator. Because of the rotation of the Earth, water piles up on the western sides of the oceans and flows toward the equator in rather narrow currents. The Gulf Stream off the coast of the United States is a familiar example, but similar currents exist in other oceans.

But both the ocean and atmospheric flows are much more complicated than those that I have described so far.

1.3   THE THERMOHALINE CIRCULATION

Consider first the ocean. Many years ago, sailors cooled their wine bottles by lowering them into the ocean far below the surface. The fact that the ocean water deep below the surface is much colder than the surface water was apparently discovered by Captain Henry Ellis of the British slave trader *Earl of Halifax*. Ellis had measured the change in temperature with depth by lowering a bucket to various depths and raising it to the surface to measure the temperature of the water. The bucket was fitted with valves that allowed water to flow through it on the way down but shutting on the way up so that water at a particular depth could be obtained for observation. Bruce Warren, in the volume *Evolution of Physical Oceanography*,[3] quoted a letter from Ellis to the Royal Society of London in 1751:

Upon the passage I made several trials with the bucket sea-gage, in the latitude 25' 13" north; longitude 25' 12" west. I charged it and let down to different depths, from 360 feet to 5346 feet; when I discovered, by a small thermometer of Fahrenheit', made by Mr. Bird, which went down in it, that the cold increased regularly, in proportion to depths, till it descended to 3900 feet: from whence the mercury in the thermometer came up at 53 degrees; and tho' I afterward sunk it to the depth of 5346 feet, that is a mile and 66 feet, it came up no lower. The warmth of the water on the surface, and that of the air, was at that time by the thermometer 84 degrees. I doubt not but that the water was a degree or two colder, when it enter'd the bucket, at the greatest depth, but in coming up had acquired some warmth.

Later in the letter, he wrote:

This experiment, which seem'd at first mere food for curiosity, became in the interim very useful to us. By its means we supplied our cold bath, and cooled our wines and water, which is vastly agreeable to us in this warm climate.

It was discovered that the deep water in the ocean far below the surface was very cold, even near the equator, where the surface was warm. This could only be true if deep ocean water was somehow coming from the cold regions near the poles. Otherwise, the warmth of the upper ocean in equatorial regions would have penetrated downward and warmed even the water near the ocean bottom. This was pointed out later by Count Rumford, who wrote in 1797:

It appears to me to be extremely difficult, if not quite impossible, to account for this degree of cold at the bottom of the sea in the torrid zone, on any other supposition than that of cold currents from the poles; and the utility of these currents in tempering the excessive heats of these climates is too evident to require any illustration.

Thus, water driven by the ocean currents at the surface finds its way to high latitudes, it loses some of its warmth to the colder atmosphere, and becomes quite cold. It also becomes saltier.

This is because when evaporation occurs at the surface only fresh water goes from the ocean surface into the atmosphere. Both increased salinity and colder temperatures make the water heavier than its surroundings, and it therefore sinks into the deep ocean, forming the cold, deep water that exists at all latitudes. Both evaporation and precipitation contribute to the salinity of seawater, of

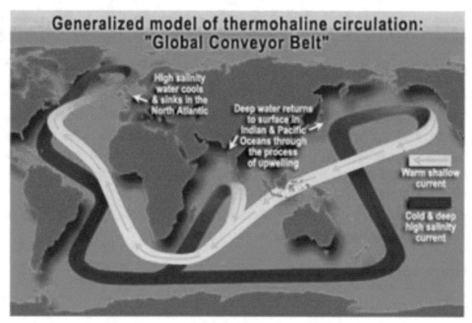

FIGURE 1.4: The thermohaline circulation: the ocean conveyor belt. The cool, salty water in the high latitudes of the North Atlantic is relatively heavy (both high salinity and cold temperatures make water relatively heavy), and this causes the water to sink into the deeper ocean. This deep water then flows into other oceans, rising to the surface and returning to the North Atlantic after completing a complex path referred to by oceanographers as the Great Conveyor Belt.[4]

course. Evaporation makes the water saltier and precipitation makes it less salty, or fresher. It happens that the North Atlantic water is saltier than Pacific Ocean water because there is a net transfer of fresh water from the Atlantic to the Pacific Ocean through atmospheric motions that carry water vapor from the Atlantic to the Pacific and deposit it in the form of precipitation. As a result, most of the deep, cold water in the ocean comes from the Atlantic, mainly from the North Atlantic. This deep water flows from the North Atlantic southward into the deep South Atlantic where it is joined by additional cold water that sinks around the coast of Antarctica. (This water sinks near the coast of Antarctica because the periodic freezing of sea ice emits very salty brine, and this heavy brine combines with the sea water to form water heavy enough to sink.) Doctor Wallace Broecker has described the "conveyor belt" that results.[4] It is illustrated in Figure 1.4. Climatologists refer to this as the thermohaline circulation. Of course, the actual flows of deep water are more complex than this, but the simplified picture in Figure 1.4 will be helpful in understanding certain features of climate change that I will discuss later.

1.4   JET STREAMS AND THE WEATHER

Now, clearly, we are speaking here of average atmospheric and oceanic motions in some sense.

There are many smaller scale motions that are very important, for example, sea breezes or the flows over mountainous terrain. Two very important features of atmospheric motions that I have not mentioned are the atmospheric jet streams. Jet streams are relatively narrow, very high velocity air currents that exist high in the atmosphere.[5]

Before about the middle of the 20th century, little was known about the details of the motion of the atmosphere at very high altitudes. During the first half of that century, humans were taking to the air in crude (by today's standards) airplanes and hot air balloons and zeppelins (blimps). But aviation really came of age during World War II. Toward the end of that war, the American Air Force prepared for a bombing mission targeting Tokyo industrial facilities, including Nakajima's Musashino plant, where a large fraction of Japan's combat aircraft engines were manufactured. On November 24, 1944, 110 B-29s took off from Saipan carrying 277.5 tons of bombs. As they neared Japan, flying at 27,000 to 30,000 ft, the winds at that altitude began to pick up. By the time they reached the target area, flying from east to west, they were fighting 140 mph headwinds. It was so difficult to gauge the drift of the bombs and other factors that most of the bombs missed their targets, and little damage was done to the Musashino plant. Later, precision bombing fared no better. These encounters with what we now recognize as an atmospheric jet stream forced the Americans to change tactics, and low-level incendiary raids replaced high-altitude missions. It is hard to believe that this was the first encounter with a phenomenon that we take for granted every day as we watch the local weather report on television.

There are actually many jet streams, the two most important to weather and climate being the subtropical jet stream and the polar front jet stream. You have seen and heard reference to the polar front jet stream during the weather portion of your local TV newscast. It happens that when two air masses of different temperatures exist closely, the wind velocity increases strongly with altitude. (This is called the thermal wind.) Thus, when the warm air from the Hadley cell meets the relatively cool air from the Ferrel cell in subtropical latitudes, an upper atmospheric jet forms: the subtropical jet (see Figure 1.2). Similarly, when the relatively warm air of the Ferrel cell meets the cold air of the polar cell, another upper air jet forms: the polar front jet. Air in these jet streams can reach very high velocities of several hundred miles per hour, flowing from west to east. Now, the polar front jet, as its name implies, forms over the frontal region between cold, high latitude air and warmer air on the equatorward side. This jet which forms at the border between cold air and warm air is not stable. Instead, it meanders around the hemisphere in a wiggly pattern, eventually giving rise to the ever-changing weather patterns that you see when you watch the daily weather report. To get a general idea of how this happens, refer to Figure 1.5. It shows a sequence of disturbances in the upper air waves. The amplitudes of the waves or wiggles increase until they break into cold regions,

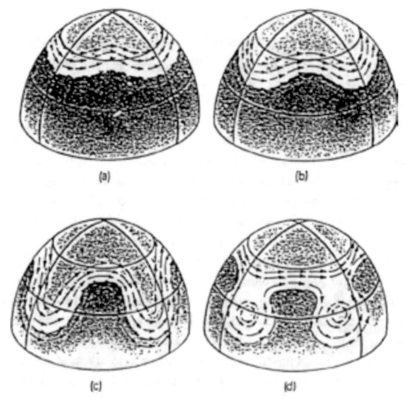

FIGURE 1.5: Upper air and the jet stream: the propagation of disturbances. The polar jet stream is not stable. This high-speed jet of air high in the atmosphere forms a wiggly pattern. The amplitude of the wiggles increases until some regions separate from the jet and form a series of regions that rotate either clockwise or counterclockwise. The regions that rotate counterclockwise contain cold air from high latitudes, whereas the regions that rotate clockwise are warm regions.[5]

which rotate counterclockwise, whereas the regions between them are warm regions, which rotate clockwise. Of course, the whole regions move toward the east because they are in the regions of the westerlies. This is clearly a very simple view of how fronts and local weather patterns form, but it gives you the general idea.

The subtropical jet stream, on the other hand, occurs along a line of descending air, or air that has a generally downward motion, that creates a high-pressure region near the ground where the diverging air currents prevent the occurrence of fronts near the ground.

But motion within the Hadley cell also has some surprises. It forms a set of several cells with longitudinal motion, that is, motion more or less perpendicular to the Hadley circulation (Figure 1.6) in the west-to-east or east-to-west direction. This has been named the Walker circulation

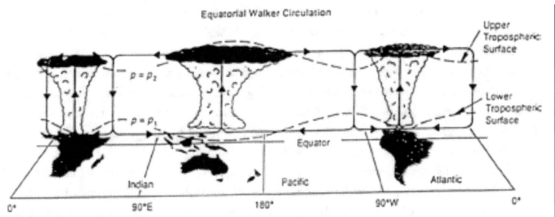

FIGURE 1.6: The Walker Circulation. The pattern of this atmospheric circulation is shown in the figure for a "normal" year; this is the La Niña condition. The wind pushes the ocean water westward off the west coast of South America. The surface ocean water that is pushed off the coast is replaced by cooler water from the deeper ocean. The rising air over the Brazilian rain forests produces clouds and rain, whereas the descending air over coastal Peru gives rise to a dry climate. When the circulation weakens or reverses, the prevailing winds no longer drive the water toward the west, and the cool water from the deep ocean no longer replaces the warm equatorial water off the coast of South America.[6]

after Sir Gilbert Walker, a scientist who postulated its existence while attempting to determine the causes of (and to predict) monsoon failures in India.

1.5   EL NIÑO AND LA NIÑA

The Walker circulation "normally" contains winds that blow westward off the west coast of South America.[6] These winds literally blow the upper ocean water off the coast. It is then replaced by water from the deeper ocean, and this water is, as we have seen, cold. Thus, the ocean surface is much warmer in the tropical western Pacific than in the tropical eastern Pacific near South America. Air above the warmer region is warm and buoyant and rises, whereas air above the cooler region sinks, maintaining the circulation that created the temperature difference. The western tropical Pacific has the warmest surface water in the entire ocean, whereas water near the west coast of South America is so cool that penguins thrive in the Galapagos Islands. The rising, moist air over the islands in that region leads to high levels of precipitation and lush tropical forests. When the Walker circulation is in the form shown in the picture, the rising air over the islands in the equatorial Pacific produces heavy rainfall, whereas the generally descending air over coastal South America leads to very dry conditions there. (In general, rising air produces rainy, wet conditions at the surface, whereas descending air produces high-pressure and dry conditions at the surface.) Further east, another region of rising air produces heavy precipitation over the Brazilian rain forests. The complex pattern over Africa leads to rainforests in some regions and very dry conditions in the Middle East. The Walker Circulation is not stable. It can and does change its intensity and direction every few years, a phenomenon known as the Southern Oscillation. When this happens, cool water no longer rises to the surface near South America. Because the region is near the equator, the water warms quickly, and the warm water spreads toward both poles. The phenomenon that I have described is, of course, the El Niño. During El Niño years, the normally very dry regions of western South America, Central America, and California become wet, and the tropical regions north of Australia become dry. The effects of El Niño go much further, affecting the weather throughout a large portion of the globe. The strong El Niño of 1998 produced damaging record rainfall in the southwestern United States, and dry conditions over the equatorial Pacific islands and in Australia have led to record forest fires because of the dryness of the forests and absence of ameliorating rain. La Niña is the opposite condition from El Niño that exists when the Walker circulation is in the condition shown in Fig. 1.6. El Niño means the "boy child" because the conditions in many cases begin around Christmas, whereas La Niña refers to the "girl child," the opposite condition. Scientists refer to the El Niño/La Niña phenomenon as an El Niño–Southern Oscillation. The Southern Oscillation refers to the changing pressure difference of the atmosphere between Tahiti and Darwin, an island off the coast of South America. When the surface pressure is higher at Darwin, the atmospheric flow is more or less like that shown in Figure 1.6. When the surface pressure is higher at Tahiti, the flow reverses, and there is an El Niño event.

The above constitutes a rather crude picture of how the atmosphere and the ocean work. I have only hinted about the difference between weather and climate, and it is very important to distinguish carefully between the two. I will now do just that.

1.6   CLIMATE AND WEATHER

As I sit in my office in New Orleans writing, this it is November, the weather outside is cold and gloomy. (Cold to someone living in New Orleans is below 50°F.) The sky is overcast. This is not "normal" for early November. Can we say that the weather in New Orleans in the fall is cold and gloomy? Of course not. The weather in New Orleans in November is "normally" beautiful, not too warm or too cool. It is one of the most pleasant months of the year. We may see 80°F days in December, but we can scarcely say that winters in New Orleans have 80°F days all the time, or even every winter. Travel guides give you the "normal" day and night temperatures so that you can pack the right clothes for a visit. But they admit that you might hit a cold or a hot spell. It is fairly safe to say, however, that winters in New Orleans are mild, and that the summers are hot and humid, whereas autumn and spring are mild and beautiful. Weather can change drastically in a period of a few hours, but the general climate of a region is not likely to change dramatically from, say, one decade to the next. We might have a relatively dry summer, but on the average over the years, south Louisiana summers are hot and humid, with considerable rainfall, mostly as afternoon showers. Weather is notoriously difficult to predict. In fact, it is essentially impossible to predict more than a week or so in advance because the solutions of the mathematical equations that are used to do computer predictions are chaotic. What this means in essence is that predictions of the weather more than a week or so from now are so sensitive to the present conditions that it would be impossible to know them (the present conditions) accurately enough to make an accurate prediction of the weather conditions more than a week hence. In predicting climate, however, we are not attempting to get all the day-to-day details but only to predict some average conditions. Will it be warm or cool in a general region in some season? Will it be rainy or dry? Is the rain likely to come as severe events, or as fewer, less severe events?

Back in the 1960s and 1970s, the comedian George Carlin had as part of his comedy act the "hippy dippy weather man." He was really a climate man. He would say that it is hot now but it should get cooler during the fall, finally getting real cold in the winter, and warming toward spring.

Think of El Niño as a sort of miniclimate change. It is really a kind of shift in the climate. The eastern Pacific Ocean has become warmer. This small change in the distribution of climate leads to dramatic changes in weather patterns around the globe.

Changes in the climate, as measured, say, by the average temperature of the Earth's surface, can produce changes in weather. This is key to your understanding of the effects of climate change on people. For example, the position and pattern of the polar jet may change, producing changes in the *distribution* of weather events. It may become dryer in some places and wetter in others. If the Earth becomes warmer, evaporation from the surface will generally increase, just as evaporation from a pot of water on your stove is greater when you turn the heat up. Because there is more evaporation, there must be higher precipitation (rainfall) over the Earth as a whole, but it may be higher in some places and lower in others, as weather patterns shift. Also, the availability of water to plants and others is the precipitation minus the evaporation, and so even if rainfall increases at a given location, if the evaporation increases, there will be less water available for use by vegetation there. So climate and weather are not the same, but they are intimately related in a sense. If the global climate changes, regional climates will change, and the weather patterns in those regions will also likely change.

Recall the discussion associated with Figure 1.5. Under what we consider "normal" conditions, the polar jet stream forms a set of wiggles that create the constantly changing weather patterns that are called fronts. Sometimes, the jet stream gets stuck in what we call a blocking pattern. During the winter of 1976–1977, such an event occurred. The jet stream normally comes across the Pacific Ocean and over the western United States near the northern states of Oregon and Washington, undulating across the U.S. mainland and creating constantly changing weather patterns. But in the winter of 1976–1977, it occupied a more or less fixed position, a blocking pattern, in which it entered across Alaska and plunged far southward across the eastern states. Because of the relatively warm air off the Pacific, Alaska had a relatively warm winter, but as the air traveled southward across Alaska and Canada, it brought cold air as far south as Florida. The southeastern states experienced one of the coldest and wettest winters in memory, whereas conditions in the western states were unusually warm and dry. I spent that winter in Oak Ridge, TN, at the Institute for Energy Analysis. Eastern Tennessee experienced more snow than the residents could ever remember. Because it seldom snows more than once or twice in Oak Ridge, and then only lightly, the city was plunged into virtual paralysis. There were no snowplows to clear the snowy streets, and people had little experience driving in the snow. The region is near the foothills of the Smoky Mountains, and the hills made driving extremely hazardous. The last snow fell at the end of March.

The average temperature in the northwestern United States was considerably higher than usual during that winter, whereas the average temperature in the southern Appalachians was several degrees colder than usual. If one lived in the northwest, one could well imagine that global warming was happening, but folks in the southeast might have seen the beginning of the next ice age! The main point here is that a single warm season in a particular region or even a few warm years in a row is not sufficient to prove global warming. We need to look for long-term global changes. But changes in climate will cause changes in weather patterns. Will a warmer climate lead to more or fewer blocking patterns and the resulting abnormal weather? We do not know.

1.7   CLIMATE CHANGES IN THE PAST

What can we learn from historical changes in the climate of the Earth? There have been periods in the past when the climate of the Earth has been much different from that of today. Between about 120,000 and about 18,000 years ago, the climate was much colder than it is in the present. The globally averaged temperature was about 4°C (6°F) colder than the present globally averaged temperature. But it was not that much colder everywhere. Equatorial temperatures were probably not much cooler than they are at present, whereas temperatures further toward the poles were very much colder. In North America, a thick ice sheet covered a large portion of the continent, reaching as far south as southern Ohio. The currently held view is that this ice age resulted from differences in the way the Earth orbits the Sun. The path of the Earth's orbit is not a pure circle, but rather an ellipse. Sometimes, the northern hemisphere summer occurs when the Earth is closest to the Sun, and sometimes it occurs when the Earth is farthest from the Sun (see Figure 1.1). (Remember, it is not the eccentricity, or the distance of the Earth from the Sun, that causes the seasons; it is the tilt of the Earth's axis.) The eccentricity of the orbit also changes with time, as does the "tilt" of the Earth's axis. Suppose that northern hemisphere summer comes when the Earth is farthest from the

Sun. This would be expected to produce milder summers. If, in addition, the tilt of the Earth's axis were small, the contrast between the seasons would be smaller, again suggesting milder summers. When these things occur together (the hypothesis goes), the snow that fell during winter would not be completely melted during the summer and would begin to accumulate year after year, resulting in the advance of ice sheets. This is referred to by climatologists as the Milankovich Theory after the Yugoslavian mathematician who was one of several people who proposed the idea.

1.8   EXPLAINING ICE AGES

Although the idea that great glacial advances had long ago occurred on Earth was generally believed by many Swiss citizens who lived in the mountains, it was only by the middle of the 19th century that it gained wide acceptance by scientists.[8] One piece of evidence was the erratic distribution of large boulders in Europe far from any possible bedrock source. Most scientists of the time preferred to believe that the boulders were moved from their place of origin by huge currents of water and mud, driven by Noah's flood described in the Old Testament. Recognizing that it was unlikely that water currents alone could move large boulders such great distances, the English geologist Charles Lyle argued that the boulders had been frozen in glaciers and that in the great flood of biblical times, boulder-laden icebergs had drifted about, depositing the boulders erratically as they melted. Charles Darwin had, in fact, reported that he had observed icebergs in the southern ocean that contained boulders.

Most geologists in the 18th and 19th centuries believed that Earth had undergone a series of catastrophes because this could explain the fossil animals that were being discovered without contradicting the word of God as presented in the Bible. Nearly every new discovery was interpreted within this context. A huge tooth measuring 6 inches long and 13 inches in diameter in a peat bog near Albany, NY, was identified as that of a human who had perished in the great flood. It was actually that of a mastodon.

One of the first people to challenge the biblical theory that the boulders were transported by Noah's flood was Louis Agassiz, a Swiss geologist whose main interest was in fossil fishes. Agassiz, who was also president of the Swiss Society of Natural Sciences, startled members at the annual meeting in Neuchatel, Switzerland, in 1837 by proposing that the erratic boulders found far from their original locations were transported by ancient glacial advances. The idea was not to easily find acceptance.

Agassiz sought to convince Reverend William Buckland, a professor of mineralogy and geology at Oxford and a widely respected geologist. But Buckland was a strong catastrophist and was dedicated to the notion that the purpose of geology was to prove that scientific findings of his day were perfectly consistent with the accounts of creation and the flood as described in the Bible. He was perfectly aware that it was unlikely that huge boulders could be transported great distances by a flood. Even the ice rafting theory proposed by Charles Lyle had difficulty explaining the boulders that were deposited high in the mountains. This would have required a sea level rise of some 5000 ft and a reasonable explanation of where so much water came from and where it went. Buckland finally and unaccountably became a convert after accompanying Agassiz on a field trip to study glacier deposits in Scotland. His conversion was at first treated by other scientists in England with scorn. It was another 20 years before the ice age theory was accepted by most geologists in England and other parts of the world.

Once the general idea was accepted, scientists around the world set out to estimate how much ice must have covered the continents during maximum glaciation. By looking for clues to how far toward the equator was the material transported by the ice sheet, one could estimate the area of ice. Clues to the thickness could be obtained if some part of it reached high up a mountain but did not reach the top. It was quickly realized that if such large ice sheets grew on land, the water must have come from the ocean, and the level of the oceans must have decreased by hundreds of feet. Scientists from Scotland and Scandinavia soon confirmed this.

By the late 19th century, evidence for a great ice age was strong, and the theory advanced by Agassiz was widely believed to be true. But now, scientists had to advance some kind of theory that explained how such an event could occur. Many were proposed. One early theory suggested that the energy radiated from the Sun was smaller. Another proposed that space dust scattered solar radiation, decreasing the solar flux that reached Earth. Some scientists were convinced that a decrease in the atmospheric loading of CO_2 was responsible. A proposal from two American scientists, Maurice Ewing and William Donn[9] (in 1956) suggested that increases in snowfall at high latitudes caused glaciers to grow in regions where it is already cold enough. If this happened, the new ice would reflect away enough sunlight to cool the region equatorward of the growing ice sheet below freezing, and the process would be self-perpetuating.

The astronomical theory of ice ages proposed by Milankovitch probably had its origin when a French mathematician named Joseph Adhemar published a book called *Revolutions of the Sea*[10] in 1842. Adhemar knew that the orbit of the Earth about the Sun was elliptical and that the rotational axis was tilted by about 23 degrees from a vertical line in the plane of the orbit. Furthermore, ancient astronomers had shown that the axis does not always point in the same direction. It wobbles, like a slowly spinning top. Adhemar reasoned that summer now occurs in the northern hemisphere when the Earth is farthest from the Sun, on the long part of the ellipse. Therefore, spring and summer seasons there experience longer periods of sunlight than does the southern hemisphere. The southern hemisphere, Adhemar believed, is currently in an ice age, which explains the presence of the Antarctic ice sheet, whereas the northern hemisphere is experiencing an interglacial. This so-called precession of the equinoxes (*equinoxes* means equal nights) occurs with a period of 22,000 years, so ice ages should occur alternately in each hemisphere every 11,000 years. But this idea was successfully challenged by the German naturalist von Humbolt, who pointed out that because the Earth is farthest from the Sun while the northern hemisphere's warm seasons were longest, each hemisphere, in fact, receives the same number of calories of solar energy each year.

Still, Adhemar's book represented notable progress in the theory of ice ages. A self-educated Scotsman named James Croll, who would plod through jobs as a mechanic, a tea-shop operator, an innkeeper, and a janitor, took up the challenge after reading Adhemar's book and became in later life a world-renowned scientist. In his book *Climate and Time*,[11] published when he was 54 years old, Croll presented a theory that took into account both the variations in eccentricity of the Earth's orbit and the precession cycle. Croll predicted that an ice age would occur when the ellipticity was largest, and the winter solstice (in one hemisphere or the other) occurred when the Earth was farthest from the Sun. Accordingly, the Earth should have been in an ice age from 250,000 years ago until about 80,000 years ago, when it entered the present interglacial period. The year after publication of his book, he was made a fellow of the Royal Society of London.

Croll's theory began to unravel when geologists discovered that the last glacial period had ended not 80,000 years ago but more nearly 10,000 to 15,000 years ago. It was to be revived by the Yugoslavian engineer and mathematician Milutin Milankovitch many years later.

Milankovitch's first contribution was the careful calculation of variations in the Earth's orbital parameters. Present scientists can only imagine the difficulty of making such calculations without the aid of modern computers. It would take Milankovitch 30 years. He calculated not only the shapes of the orbital parameters but also the distribution and variation of the radiation reaching the planet. Although Adhemar and Croll were convinced that reduced radiation at high latitudes during winter were the cause of ice ages, Milankovitch was not convinced and asked the great German climatologist Wladimir Koppen for his opinion. Koppen pointed out that reduced winter radiation at high latitudes could not strongly affect snowfall because the temperature at high latitudes was always cold enough for snow to accumulate. During the summer, however, glaciers tend to melt. Therefore, if summers were colder than usual, the accumulated snow would not all melt, and glaciers would gradually grow. This is the modern version of the Milankovitch theory that is generally accepted by climatologists. But there was another ace in the deck.

1.9 CARBON DIOXIDE AND CLIMATES PAST

In 1980, an article by R.J. Delmas and others entitled "Polar Ice Evidence that Atmospheric CO_2 20,000 yr BP was 50% of Present" appeared in *Nature*.[12] The evidence that these scientists presented was the CO_2 content of air bubbles buried deep in polar ice. This is presumed to be atmospheric air trapped in the snow at about the time the snow was deposited. As more snow builds up, older snow is compacted and pushed deeper in the ice. Later in that decade, a group of Russian, French, and American scientists collaborated to study the concentration of various gases trapped in air pockets buried deep in the polar ice cap in central Antarctica at the Russian Vostok station. French scientists at the glaciological laboratory at Grenoble showed that the atmospheric concentration of CO_2 was significantly lower (by 25–30%) during the entire 100,000 year glacial cycle. Scientists are not certain why this happened, but it may point to an important "feedback" that is inherent in the climate system. Colder ocean water would probably absorb CO_2 from the atmosphere more readily than warmer water. It is possible that as the climate cooled because of changes in the Earth's orbit as suggested by Croll and Milankovitch, the ocean absorbed some atmospheric CO_2, causing the Earth to cool more, and then repeating the cycle.

So we know now that the CO_2 concentration in the atmosphere during this ice age was far below the recent preindustrial concentration. Accordingly, the lower concentration of this greenhouse gas should have contributed to the cooling during that period. There have also been periods during the distant past when the atmospheric CO_2 concentration was much larger than it is today. During the mid-Cretaceous period, about 100 million years ago, the atmosphere contained some 5 to 10 times the present amount (note the large uncertainty). The Earth was warmer by about 10°C (about 15°F) on average. Once again, the increase in temperature was concentrated at high latitudes; that is, the equatorial temperatures were only slightly warmer than they are today, but the regions nearer the poles were very much warmer, so much so that there was no polar ice, and sea levels were much higher than they are today. By knowing these two climatic extremes, Doctor Martin Hoffert of New York University and Doctor Curt Covey of Lawrence Livermore National Laboratory were able to estimate that a doubling of atmospheric CO_2 would raise the Earth's average temperature by about 2.5°C, or about 4°F.[13] As we will see, this is about the average answer that climatologists predict using large computer models of the climate.

In the next chapter, I will discuss the evidence that greenhouse gases, most importantly, CO_2, are building up in the atmosphere and explain why this is happening. Then, in the next chapter, I will explain how the so-called greenhouse effect works.

NOTES AND REFERENCES

1. Cole, F.W., *Introduction to Meteorology*, Wiley, New York, 1970

2. Crowley, T.L., and North, G.R., *Paleoclimatology*, Oxford University Press, Oxford, U.K., 1991.

3. Warren, B.A., "Deep Circulation of the World Ocean," in *Evolution of Physical Oceanography*, Warren, B.A., and Wunsch, C., Eds., MIT Press, Cambridge, MA, 1981.

4. Broecker, W.S., Andree, M., Wolfli, W., Oeschger, H., Bonani, G., Kenneth, J., and Peteet, D.: "The Chronology of the Last Deglaciation: Implications to the Cause of the Younger Dryas Event," *Paleoceanography*, vol. 3, pp. 1–19, 1988.

5. Reiter, E.R.F, *Jet Streams: How Do They Affect our Weather?*, Doubleday and Company, Norfolk, U.K., 1967.

6. Bjerkness, J., "Atmospheric teleconnections from the equatorial Pacific," *Monthly Weather Review*, vol. 97, pp. 163–172, 1969.

7. Crowley, T.L., and North, G.R., *Paleoclimatology*, Oxford University Press, Oxford, U.K., 1991.

8. Imbrie, J., and Imbrie, K.P., *Ice Ages: Solving the Mystery*, Enslow Publishers, Berkely Heights, N.L., 1979. Most of the discussion in this section comes from this fascinating historical account.

9. Ewing, M., and Donn, W.L., "A theory of ice ages," *Science*, vol. 123, pp. 1061–1066, 1956.

10. Adhemar, J.A., *Revolutions de la mer*, privately published, Paris, 1842.

11. Croll, J., *Climate and time*, Appleton & Co., New York, 1875.

12. Delmas, R.J., Ascensio, J.M., and Legrand, M., "Polar ice evidence that atmospheric CO_2 20,000 yr BP was 50% of present," *Nature*, vol. 284, March, 1980, pp. 155–157.

13. Hoffert, M.I., and Covey, C., "Deriving global climate sensitivity from paleoclimate reconstructions," *Nature*, vol. 360, pp. 573–576, 1992.

• • • •

CHAPTER 2

Are the Concentrations of Greenhouse Gases in the Atmosphere Increasing?

The atmosphere of Earth consists of a mixture of gases. The composition of dry air up to a height of about 50 mi is remarkably homogeneous. It is about 78% nitrogen and about 21% oxygen. The remaining 1% is composed of many so-called trace gases, among the most important of which are CO_2 and methane (CH_4). Water vapor is also present, of course, and the amount varies widely. It is nearly 0% in desert regions and perhaps 4% in the warm, humid tropics. CO_2 makes up only about 0.03% of the atmospheric volume, about 300 parts for each million parts of the atmosphere, whereas methane makes up only about 0.00017 % of the atmosphere's volume. We usually speak of these values as parts per million by volume (ppmv). These are very small concentrations, but these and other trace gases, as well as water vapor interact very strongly with radiation, as we shall see, and so they are very important to the heat balance of Earth and, thus, to its climate. They are greenhouse gases. Greenhouse gases are those gases that absorb radiation from relatively low temperatures, such as the temperatures of the Earth and its atmosphere, but do not absorb nearly so strongly high-temperature radiation such as that coming from the Sun.

The atmospheric concentrations of a number of so-called greenhouse gases (including CO_2 and CH_4) have been increasing during the past 150 or so years. How do we know this? Very careful measurements have been made at a number of locations around the world since the middle of the 20th century. Figure 2.1 shows how CO_2 has increased during the period 1750–2000.

In 1955, systematic and careful measurements of the CO_2 content of the atmosphere by Doctor Charles D. Keeling began in Hawaii and at the South Pole.[1] They show that the CO_2 content of the atmosphere rose from about 315 ppmv in 1958 to the more than 366 ppmv by the late 1980s. The IPCC Fourth Assessment[2] reports the value as 379 ppmv in 2005. Before these measurements began, we know from past measurements[3] of air trapped in polar ice that the level of CO_2 remained approximately constant at about 280 ppmv from 1000 until 1800.

2.1   EARLY IDEAS ABOUT CARBON DIOXIDE

A number of scientists realized that the presence of CO_2 and water vapor in the atmosphere enhanced the warmth of the Earth by allowing solar radiation to penetrate the atmosphere while

trapping the lower temperature given off by the Earth. As far back as 1824, the great French mathematician Fourier[1] had described the greenhouse effect of Earth's atmosphere, comparing it with a
glass covering a container.

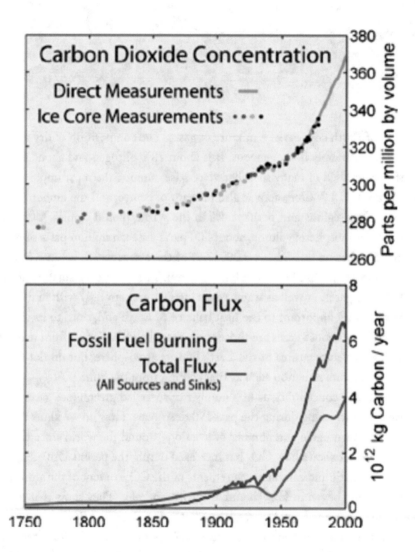

FIGURE 2.1: The top panel shows the increase in CO_2 between 1750 and 2000. The dots are data from
measurements of CO_2 in gas bubbles from ice cores, and the blue line represents smoothed data from
instruments. The bottom panels shows the flux to the atmosphere (blue) and the amount that stays in
the atmosphere after absorption by the biota and the oceans. The net flux of carbon into the atmosphere
clearly tracks closely the flux emitted by burning fossil fuels.

The British physicist John Tyndale published an article[5] in 1861 stating that the amount of CO_2 in the atmosphere could have a strong effect on climate. Tyndale's interest, however, was in challenging the ice age theory advanced by Croll, recognizing that carbon was cycled through plants and perhaps the ocean and was therefore subject to variations over time. Some 35 years after Tyndale's article appeared, the Swedish chemist Svante Arrhenius in 1896 stated that "the selective absorption of the atmosphere is, according to the researches of Tyndale, Lecher and Pernter, Rontgen, Heine, Langley, Angstrom, Paschen, and others, of a wholly different kind. It is not exerted by the chief mass of air, but in a high degree by aqueous vapor and carbonic acid, which are present in the air in small quantities."[6] Arrhenius was also interested in refuting Croll's theory of ice ages, but he also realized the other problem: the one that would probably arise with too much CO_2. He wrote that, "We are evaporating our coal mines into the air," and the result might be a warming of Earth. In this remarkable article, Arrhenius considered both water vapor and ice albedo feedback (which I will discuss in Chapter 3) and predicted that the global temperature would increase by 5°C to 6°C if the CO_2 in the atmosphere were doubled. The American geologist Thomas Chamberlin, again presenting an alternative to Croll's theory, referred to CO_2 changes as a possible cause.[7] He went on to refer in passing to the possible effect on climate of the continued use of fossil fuels in cities, with the attendant production and emission into the atmosphere of CO_2.

Over the next four decades, there appears to have been little interest in atmospheric CO_2. But in a series of articles published between 1938 and 1961, a British coal mining engineer named George Callendar suggested again that fossil fuel burning could increase the atmospheric loading of CO_2 enough to change the climate.[8] By analyzing the previous hundred years of CO_2 measurements, he found hints of an increase. By estimating the amount of CO_2 produced, he noted that the increase in the atmosphere compared well with that produced by fossil fuel burning. There is, however, no indication of how he estimated total fossil fuel burning, and it appears that he assumed that all the CO_2 resulting from fossil fuel burning stays in the atmosphere. Using data from the World Weather Records of the Smithsonian Institution, he estimated the increase in global temperature, which he attributed to increased CO_2. He regarded the increase in temperature as beneficial to mankind and stated that, "there is no danger that the amount of CO_2 in the air will become uncomfortably large because as soon as the excess pressure becomes appreciable…the sea will be able to absorb this gas as fast as it is likely to be produced." Similar views were expressed by other scientists at the time. Roger Revelle, at the time the director of the Scripps Institution of Oceanography, collaborated on an article in 1957 with Hans Seuss, in which they concluded that most of the CO_2 produced by fossil fuel burning since the beginning of the industrial revolution must have been absorbed by the oceans.[9]

The data reported by Callendar must have finally convinced the scientific community of the importance of making accurate measurements. During a conference on atmospheric chemistry at the University of Stockholm in May 1954, a Swedish scientist, Doctor Kurt Buch, proposed establishing a network of sampling stations in Scandinavia. The network began operating in November 1954, reporting data regularly to the journal *Tellus*. In their first report, the scientists suggested that it would be highly desirable if similar measurements were made in other locations.

2.2   THE KEELING CURVE

Rather crude measurements had been made for quite some time, but when Charles D. Keeling was given the problem, he set out to make very accurate measurements.[10] Keeling had received a PhD in chemistry from Northwestern University and was looking for a postdoctoral position when he was hired by Doctor Harrison Brown at the California Institute of Technology. Keeling was an enthusiastic outdoorsman and enjoyed camping and backpacking. He was looking for a position that would allow him to work outdoors, so when he heard Brown et al. discussing whether the CO_2 dissolved in water bodies was in balance with that in the air above the water, he volunteered to conduct an experiment to find out whether the CO_2 in the air was in equilibrium with that in water bodies. Finding no really accurate device on the market for measuring CO_2 in the air, he set out to invent one. Actually, he set out to vastly improve the accuracy of an existing manometric device. The amount of CO_2 in the atmosphere is actually very small, about 3/100 of 1% of the volume of the atmosphere. Keeling's device had to be very accurate indeed to detect subtle differences between the concentrations in water and air.

It took more than a year for Keeling to perfect his device. In 1955, he collected air on the rooftop of Mudd Hall on the Caltech campus and determined the CO_2 concentration to be 310 ppmv. (This means that the fraction of the atmosphere that was CO_2 was 310/1,000,000, or 0.031%.) Subsequently he made measurements in Yosemite National Park and in the Cascade Mountains with similar results. Soon after that, Doctor Oliver Wulf of the Weather Bureau suggested that Keeling contact the head of the Weather Bureau's Office of Meteorological Research, Harry Wexler, to consider monitoring background concentrations of atmospheric CO_2. The International Geophysical Year had been in the planning stages for years and was about to begin. Scientists from some 70 counties around the world planned to spend 18 months observing and measuring the state of the planet and its oceans and atmosphere. Roger Revelle was one of the planners of the International Geophysical Year, and he was now wondering whether the atmospheric content of CO_2 was rising. Wexler met with Keeling in his office in Washington, DC, where he was asked whether and how he could determine if atmospheric CO_2 was increasing. Keeling proposed measuring the atmospheric CO_2 content in many places around the world. They immediately agreed on a program at the Mauna Loa Observatory in Hawaii.

Mauna Loa would be a perfect location for a weather station because it was distant from sources of industrial pollution yet accessible from large cities. A small wood building had been constructed in 1951 about 280 ft below the summit at the 13,400-foot level. This observatory closed in

1954 because of difficulty in maintaining the access road. However, in 1955, the National Bureau of Standards provided funds to build the larger permanent structure that remains there today.

Roger Revelle was a proponent of periodically sampling the atmosphere for CO_2 by aircraft rather than from fixed surface stations. Such measurements were made along with measurements from Mauna Loa, and Ben Harlan of the Weather service also established a station in Antarctica.

When the first measurements were made from Mauna Loa, they turned out to be 311 ppmv, in agreement with Keeling's previous measurements and with those of several other scientists. It was thought that this showed that the atmospheric CO_2 content remained constant with time. Subsequent data showed very soon how unjustified was this initial reaction. During the first year, concentrations climbed to 315 ppmv during the winter, returning to near 311 ppmv toward the end of the summer. This was quickly recognized as a seasonal cycle. During the spring and summer, plants remove CO_2 from the atmosphere by photosynthesis as they grow. During the winter, it is returned to the atmosphere as plants shed their leaves and die back. Measurements have now continued for more than 50 years, with many new sites being established in the intervening years.

Figure 2.2 shows a more detailed graph of the atmospheric CO_2 concentrations measured at Mauna Loa. Figure 2.1 does not show the seasonal cycle because the data plotted are annual averages.

Measurements are now made from pole to pole, from the South Pole to Point Barrow, AK. Most stations now measure the CO_2 concentration continuously using infrared gas analyzers. They all show the same increasing trend. When someone asked Keeling in the 1960s why he was measuring

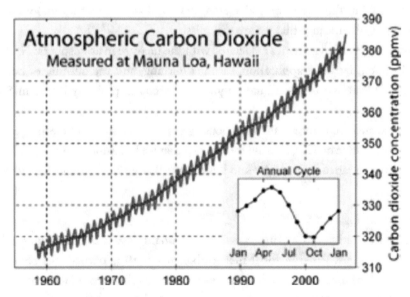

FIGURE 2.2: The Keeling curve showing the increase in atmospheric CO_2 since his careful measurements began. The inset shows the seasonal cycle.

atmospheric CO_2, he reportedly answered, "You mean what does it mean for the man in the street?" Keeling's measurements were begun as a result of pure curiosity by a very talented scientist. As we will soon see, the result is one of the most important sets of data gathered in the 20th century.

2.3   OTHER GREENHOUSE GASES

The atmospheric concentrations of other important greenhouse gases have also been increasing.[11] The concentration of methane, which was fairly constant from 1000 to 1800, has doubled since 1800. Nitrous oxide (N_2O) has also increased since 1800, being nearly constant back to 1000. Why has the atmospheric concentration of methane increased? This we do not know for sure. Rice paddies emit methane into the atmosphere. The increase in rice agriculture has probably contributed. Cattle and other ruminant animals produce methane in their digestive systems and emit it to the atmosphere. Some may be coming from natural wetlands and from landfills. Approximately one quarter of methane emissions probably comes from the production of energy, directly or indirectly, through coal mining and the transmission, distribution, venting, and flaring of natural gas. As an interesting historical note, the *Study of Man's Impact on Climate*, published by the Massachusetts Institute of Technology Press in 1971, stated categorically that "because CH_4 (methane) has no direct effect on climate or the biosphere it is considered to be of no importance to this report."[12] Four years later, the World Meteorological Organization's report on climate change came to the same conclusion. Methane has since been discovered to be a powerful greenhouse gas.[13] However, methane emissions are currently estimated to contribute about 25% of the radiative forcing responsible for climate change, but the initial misunderstanding (i.e., that it was of no importance) indicates that there might be other things that we are still missing.

The dominant human activities associated with the increase in atmospheric N_2O are thought to be soil cultivation, fertilizer application, savanna burning, and the burning of both wood and fossil fuel. N_2O is expected to contribute only a small amount, probably less than 5%, to global warming.[13]

Thus, there are several important greenhouse gases whose atmospheric concentrations are increasing. I will concentrate on CO_2 because it is the one that we know for certain is produced by burning fossil fuels—by human use of energy—and it is the one that produces most of the global warming.

2.4   THE ENERGY CONNECTION

Why has the concentration of CO_2 increased? The standard answer is the burning of fossil fuels: coal, oil, and natural gas. When a fossil fuel (carbon material) is burned, the chemical reaction (burning, or oxidation) involved produces CO_2, which is generally released into the atmosphere. In 1973, Keeling described a systematic method of estimating the amount of CO_2 emitted from fossil fuel consumption, using information from the United Nations Department of International Economic and Social Affairs.[14] His methods have been refined and improved by Gregg Marland and the late Ralph Rotty of the Institute for Energy Analysis in Oak Ridge, TN. They examined fuel use by the type of fuel and how much each fuel contributed to the emission of CO_2 to the atmosphere. Rotty and Marland included CO_2 emissions from cement production and the flaring of natural gas. Keeling estimated that in 1860, 93.3 million metric tons of carbon in the form of CO_2 was emitted to the atmosphere from burning fossil fuels, almost all from burning coal. Estimates in 1984 by Rotty and Marland indicated that the number had risen to more than 6000 million metric tons each year (6 Gt/yr; 1 Gt = 1012 kg).[15] The IPCC Fourth Assessment Report reports that the value had risen to approximately 7.2 Gt/yr in 2005. In a testimony before Congress, Gregg Marland put the 2005 figure (from fossil fuels and cement manufacture) at 7.85 Gt. The CO_2 flux entering the atmosphere and the rate of emission of CO_2 from fossil fuel burning increase together, but this alone does not prove that the two are related. It might be a coincidence. Such things happen in science and in nature. But there are ways to tell.

Many parts of the Earth system contain carbon and exchange it with the atmosphere. For example, plants (biota) and soils contain carbon in amounts comparable to that contained in the atmosphere, and the oceans contain many times more. The estimated amounts of these reservoirs are shown in Figure 2.3. The amounts are measured in gigatons carbon (1 Gt = 10^{15} g, or 1,000,000,000,000,000 g). The atmosphere currently contains about 750 Gt C, compared with 550 Gt C in plants, 1500 Gt C in soils, 1000 Gt C in the upper layer of the ocean, and some 38,000 Gt C in the deep ocean. Fossil fuel reserves (coal, oil, and natural gas yet to be mined) contain 5000 to 10,000 Gt C. Annual exchange rates of carbon between these reservoirs are very large. The exchange rate between the ocean and the atmosphere is estimated at 90 Gt C/yr and that between the terrestrial biosphere and the atmosphere is estimated at 100 Gt C/yr. For comparison, fossil fuel use currently emits about 7 to 8 Gt C/yr, whereas deforestation is estimated to emit between 1 and 2 Gt C/yr. Of this, only about one half shows up as atmospheric increase as indicated in Figure 2.1. The other half is probably absorbed by the ocean, soils, and plants.

Obviously, atmospheric CO_2 levels are maintained by a delicate balance of large flows between large reservoirs. If the flow of carbon between these large reservoirs is so large, the natural question (which has been raised by some) is, "How can we be certain that the increase in atmospheric CO_2 is not simply a result of natural processes?" It is a legitimate question. Some of the evidence that the atmosphere's increase in CO_2 is the result of fossil fuel burning is circumstantial. One bit of circumstantial evidence comes from the similarities (even in details) of the carbon emission increase data and the data on the increasing atmospheric CO_2. For example, the rate of growth of atmospheric CO_2 shows slight breaks in 1973 (corresponding to the oil embargo by the Organization of Petroleum Exporting Countries) and in 1979 (corresponding to the Iran–Iraq war). During these periods,

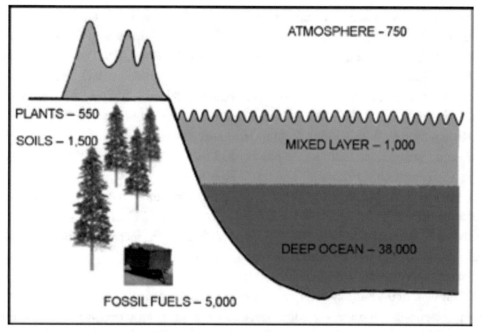

FIGURE 2.3: Carbon reservoirs. The amounts of carbon (given in gigatons, or 10^{15} g C) that is stored in plants, soils, ocean, and unused carbon fuels still buried in the Earth are much greater than that in the atmosphere. The exchange rates (gigatons carbon per year) were approximately in equilibrium before the industrial revolution, as demonstrated by the fact that the atmospheric CO_2 content of the atmosphere remained constant from the last ice age until the industrial revolution, when humans began burning fossil fuels.

the rate at which fossil fuel was being burned decreased, and so the rate at which atmospheric CO_2 was building up in the atmosphere was smaller. Another piece of circumstantial evidence comes from the fact that the balance between the reservoirs must have been remarkably stable between 1000 and 1800, when the atmospheric levels of CO_2 remained stable at a level of about 280 ppmv. It is difficult to reconcile this apparent stability (and the implied equilibrium between flows of carbon into and out of the large reservoirs) with the fact that just when fossil fuel burning began to rapidly increase, the atmospheric CO_2 concentration began to increase at an unprecedented rate.

The difference between the CO_2 concentrations in the northern and southern hemispheres has increased from 1 to 3 ppmv during the last 30 years, higher in the northern hemisphere, where most fossil fuel CO_2 is emitted, another bit of circumstantial evidence.[16]

The most compelling piece of evidence, however, is associated with the difference between carbon isotopes, chemically identical forms of carbon but with different atomic weights.[16] There are three common isotopes of carbon: ^{12}C, ^{13}C, and ^{14}C. Nearly 99% of ordinary carbon is in the form ^{12}C, with ^{13}C making up most of the remainder along with a very small amount of the radioactive isotope [14]C. When plants grow, they selectively take up [12]C. Because fossil fuels were originally plant and animal matter, they also contain proportionally less ^{13}C than is present in the atmosphere. Because fossil fuel has been stored in the Earth for much longer than it takes the radioactive isotope ^{14}C to decay, it contains none of this isotope. CO_2 emitted from burning of coal, oil, and natural gas has less ^{13}C than atmospheric carbon. Therefore, as CO_2 enters the atmosphere from burning of fossil fuels, the ratio ^{13}C/^{12}C should decrease over time. The changes can easily be measured by modern mass spectroscopy. The lower panel of Figure 2.4 shows the decrease in the isotope [13]C (from Keeling et al.[17]) alongside the smoothed CO_2 emissions curves from Marland et al.[18] The symbol $\delta^{13}C$ means the change in the ^{13}C isotope, and note that it is negative. The upper two curves in this figure illustrate the atmospheric CO_2 content as measured at the Mauna Loa station and also measurements by Manning and Keeling of the atmospheric O_2 (oxygen) content.[19] The fact that the O_2 content is decreasing

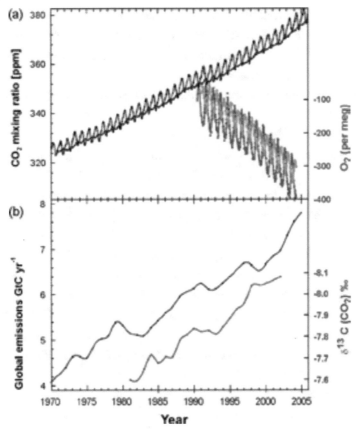

FIGURE 2.4: The lower panel shows the smoothed global emissions of CO_2 (in black) alongside the change in the carbon isotope ^{13}C in the atmosphere (in red). The upper panel shows the Keeling curve again alongside the decreasing oxygen content (two locations, one in red and one in blue).

simply demonstrates that some of the CO_2 is being absorbed in the ocean. This is important because it changes the chemistry of the ocean. I will discuss this important fact in Chapter 5, where the consequences of global change will be presented.

Thus, there is a compelling case for concluding that the buildup of CO_2 in the atmosphere is caused by burning fossil fuel. The main greenhouse gas that is increasing rapidly is CO_2, and we know with certainty where it is coming from. It is coming mostly from the burning of fossil fuels to satisfy the energy requirements of humans.

NOTES AND REFERENCES

1. Keeling, C.D., "The concentrations and isotopic abundances of carbon dioxide in the atmosphere," *Tellus*, vol. 12, 1960.

2. The web site of Carbon Dioxide Information Analysis Center (CDIAC) at Oak Ridge National Laboratory has a world of information on CO_2 and other radiatively active gases, as well as many other data of interest: http://cdiac.esd.ornl.gov/.

3. Siegenthaler, U., Friedli, H., Loetscher, H., Moore, E., Neftal, A., Oeschger, H., and Stauffer, B. "Stable-isotope ratios and concentrations of CO_2 in air from polar ice cores," *Annals of Glaciology.*, vol. 10, pp. 1–6, 1988.

4. Fourier, J., "Memoire sur les Temperatures du Globe Terrestre," *Annals de Chimis est de Physique*, vol. 27, pp. 136–167, 1824.

5. Tyndale, J., *Philosophical Magazine, Journal of Science.*, vol. 22, pp. 169–194, 1861.

6. Arrhenius, S., "On the influence of carbonic acid in the air upon the temperature of the ground," *Philosophical Magazine*, vol. 41, pp. 237–276, 1896.

7. Chamberlin, R.C., "An attempt to frame a working hypothesis on the cause of glacial periods on an atmospheric basis," *Journal of Geology*, vol. 7, pp. 576, 667, and 751, 1899.

8. Callendar, G.S., "The artificial production of carbon dioxide and its influence on temperature," *Quarterly Journal of the Royal Meteorological Society*, vol. 64, pp. 223–237, 1938; Callendar, G.S., "Variations of the amount of carbon dioxide in different air currents," *Quarterly Journal of the Royal Meteorological Society*, vol. 66, pp. 395–410, 1940; Callendar, G.S., "Can carbon dioxide influence climate?," *Weather*, vol. 4, pp. 310–314, 1949; Callendar, G.S., "On the amount of carbon dioxide in the atmosphere," *Tellus*, vol. 10, pp. 243–248, 1958; Callendar, G.S., "Temperature fluctuations and trends over the earth," *Quarterly Journal of the Royal Meteorological Society*, vol. 87, pp. 1–12, 1061.

9. Revelle, R.H., and Seuss, H.E., "Carbon dioxide exchange between atmosphere and ocean and the question of an increase of atmospheric CO_2 during the past decades," *Tellus*, vol. 9, pp. 18–27, 1957.

10. Weiner, J., *The Next Hundred Years*, Bantam Books, 1990.

11. Wuebbles, D.J., et al., "Emissions and budgets of radiatively important atmospheric constituents," in *Engineering Response to Global Climate Change*, R.G. Watts, Ed., Lewis Publishers, 1997.

12. SMIC, *Inadvertent Climate Modification: Report of the Study of Man's Impact on Climate*, MIT Press, Cambridge, MA, 1971.

13. Ramanathan, V., Cicerone, R.J., Singh, H.B., and Kiehl, J.T., "Trace gas trends and their potential role in climate change," *Journal of Geophysical Research*, vol. 90, pp. 5547–5566, 1985.

14. Keeling, C.D., "Industrial production of carbon dioxide from fossil fuels and limestone," *Tellus*, vol. 25, pp. 174–198, 1973.

15. Rotty, R.M., and Marland, G., "carbon dioxide emissions from fossil fuels: a procedure for estimation and results for 1950–1982," *Tellus*, vol. 36B, pp. 232–261, 1984.

16. Houghton, J., *Global Warming*, 2nd ed., Cambridge University Press, 1997.

17. Keeling, C.D., Bollenbacher, A.F., and Whorf, T.P., "Monthly atmospheric 13C/12C isotope ratios for 10 SIO stations," In *Trends: A Compendium of Data on Global Change*, Carbon Dioxide Information Analysis Center, Oak Ridge National Laboratory, U.S. Department of Energy, Oak Ridge, TN, 2005, http://cdiac.ornl.gov/trends/co2/iso-sio/iso-sio.html.

18. G., Marland, Boden, T.A., and Andres, R.J., "Global, regional and national CO_2 emissions," In *Trends: A Compendium of Data on Global Change*, Carbon Dioxide Information Analysis Center, Oak Ridge National Laboratory, U.S. Department of Energy, Oak Ridge, TN, 2006.

19. Manning, A.C., and Keeling, R.F., "Global oceanic and land biotic carbon sinks from the scripps atmospheric oxygen flask sampling network," *Tellus*, vol. 55B, pp. 95–116, 2006.

• • • •

CHAPTER 3

The Greenhouse Effect and the Evidence of Global Warming

3.1 THE GLOBAL ENERGY BALANCE

The Earth is heated by radiation from the Sun. In turn, an equal amount of radiation must leave the top of the atmosphere for the Earth to maintain a balance so that it does not either heat up or cool down. The interaction of radiation with the atmosphere and the Earth's surface is very complicated. However, we can get a general picture of it by summarizing the Earth–atmosphere energy balance on average for the whole globe (i.e., global average; Figure 3.1).[1] Suppose that 342 W/m² of radiant energy impinge upon the top of the atmosphere, about the power of three and a half 100-W light bulbs. Not all of it reaches the surface of the Earth. About 67 W/m² is absorbed by the atmosphere, mostly by ozone but some by water vapor and dust or other particulate pollution. Another 77 W/m² is reflected by clouds and aerosols (small particles in the atmosphere). The Earth's surface reflects 30 W/m² back to space. The remaining 168 W/m² is absorbed by the surfaces of the land and the oceans. The surface of the Earth (land and oceans) directly heats the atmosphere (sensible heat), losing about 24 W/m², whereas evaporation from the surface (latent heat) accounts for about 78 W/m². The Earth's surface and the atmosphere emit radiation that has different characteristics from those of solar radiation. I will say more about that shortly. About 390 W/m² of this "long-wave" radiation is emitted by the surface. Almost all of this is absorbed by clouds, water vapor, CO_2, and several other gases. Forty watts per square meter escapes through the atmosphere. About 324 W/m² is reemitted downward to the surface. The atmosphere, together with clouds, emits another 195 W/m² into space.

Notice that the sum of the energy units must balance both at the surface and at the top of the atmosphere. The sum of the energy passing into the atmosphere from the Sun must be equal to the amount of radiant energy leaving the top of the atmosphere, which is the solar radiation reflected back plus the long-wave radiation leaving from the top of the atmosphere. Otherwise, the Earth–atmosphere would heat up. The same is true at the surface. The 168 W/m² of solar energy absorbed plus the 324 W/m² of long-wave heating from the atmosphere must equal 390 W/m² of long-wave radiation emitted by the surface plus the sensible heat (24 W/m²) plus the latent heat leaving through evaporation (78 W/m²). Interrupting any of the processes can cause the system to be "out of balance." For example, if the radiation from the Sun were to increase to, say, 350 W/m², the radiation escaping from the top of the atmosphere would have to increase by 8 W/m² to bring the system back into equilibrium. Many things could happen to cause the radiation loss from the top of the atmosphere to increase. For example, the surface and the atmosphere could become warmer, and because the radiation increases with temperature, the long-wave radiation leaving the top of the atmosphere would increase. If that happened, there would probably be less ice and snow on the warmer Earth surface. Ice and snow are excellent reflectors of solar radiation, and so, less radiation would be reflected from the surface, reducing the 30 W/m² of solar radiation leaving the atmosphere because of reflection from the surface, and again putting the system out of balance.

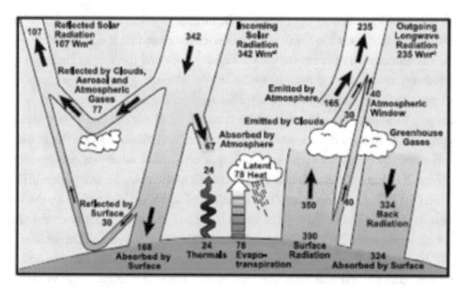

FIGURE 3.1: The Earth–atmosphere global energy balance. If neither the atmosphere nor the surface of the Earth (including land and sea areas) is changing in temperature, then the net amount of heat entering and leaving both the atmosphere and the land and sea areas must be zero. This figure shows the pathways of 342 W/m² of solar radiation bearing down on the top of the atmosphere. About 168 W/m² of this very high temperature (or shortwave) solar radiation is absorbed by the Earth. The rest is either reflected back out of the atmosphere or absorbed by the atmosphere. In addition, about 324 W/m² of long-wave radiation from the atmosphere and clouds is absorbed by the surface. The total (168 + 324 = 492 W/m²) must now leave the surface through sensible heat (caused by the temperature difference between the Earth and the atmosphere), latent heat (caused by evaporation), and long-wave radiation emitted by the surface. Most of the long-wave radiation is captured by the atmosphere, which emits about 235 W/m² out of the top of the atmosphere. If the amounts of water vapor and CO_2 in the atmosphere increase, more of the infrared radiation will be absorbed within the atmosphere, and the atmosphere and Earth must compensate by emitting more radiation. To do this, they must become warmer.

3.2   FEEDBACKS

This means that the surface and the atmosphere would have to warm even more for the long-wave radiation to compensate for the increase in solar radiation absorbed by the surface. This is called a positive feedback because it reinforces the warming initiated by the increase in solar radiation. I will discuss this feedback in some detail shortly.[2] Note now that if the solar radiation increased and the atmosphere warmed, the temperature of the surface must increase. Otherwise, the heat budget at the surface would be out of balance. To compensate, in addition to less solar radiation being absorbed (the positive feedback mentioned above), the latent heat will likely increase; that is, more evaporation will occur at the surface. This means that the water vapor content of the atmosphere would increase. Water vapor is a strong greenhouse gas, and so an increase in water vapor in the atmosphere would increase the absorption of long-wave radiation and its return to the surface (the 324 W/m^2 of long-wave radiation to the surface shown in Figure 3.1). This means that the surface must heat up even more, and so it is another positive feedback. As we will see below, there are also likely to be negative feedback, which ameliorate warming or cooling.

The greenhouse effect is caused by the interruption of part of the system wherein the long-wave radiation is absorbed by atmospheric gases and clouds. Radiation occurs as waves. The higher the temperature of the emitting source, the higher the frequency and the shorter the wavelength of the radiation. Radiation coming from the Sun is short-wave radiation. Radiation emitted from the Earth or its atmosphere is long-wave radiation. Molecules can absorb and reemit radiation, and they do so selectively according to the wavelength of the radiation.[3] CO_2, water vapor, methane, and a number of other gases either do not absorb the short-wave radiation characteristic of radiation from the Sun or do so very weakly. However, these gases absorb long-wave radiation very strongly. This means that an increase in the atmospheric content of CO_2, for example, would throw the system out of balance by increasing the absorption by the atmosphere of the long-wave radiation emitted from the surface. The atmosphere would then become warmer, increasing the Back Radiation, and, consequently, the net long-wave absorption at the surface (324 W/m^2 in Figure 3.1) would increase. But this brings the surface out of balance. To bring the surface back into heat balance, the long-wave radiation, the sensible heat, and the latent heat leaving the surface will all increase. This can only be accomplished by a warming of the surface. Now, the increase in latent heat will increase the moisture content of the atmosphere (warmer air is able to hold more moisture), and because water vapor is a strong greenhouse gas, the long-wave absorptions and emissions depicted in Figure 3.1 will all increase again. This is illustrated in Figure 3.2. The area of ice and snow will also decrease when the surface warms, decreasing the 30 W/m^2 of solar radiation reflected by the surface. This means that the 168 W/m^2 of solar radiation absorbed will increase, again requiring a further increase in surface and atmospheric temperatures to bring the surface back into heat balance.

But now things get just a bit more uncertain. Surely a changed surface temperature will cause a change in vegetation patterns, and this will no doubt change the way the surface (apart from snow

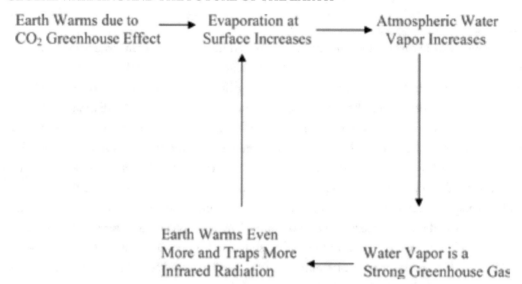

FIGURE 3.2: An illustration of the water vapor feedback. When the surface of the ocean warms because of CO_2 greenhouse effect, more evaporation occurs. Also, a warmer atmosphere can hold more water vapor. Water vapor is itself a very strong greenhouse gas. Thus, the increase in atmospheric water vapor results in a further increase in global temperature. This effect is called a positive feedback.

and ice) reflects solar radiation. Cloud patterns will probably also change, and herein lies one of the most important "jokers" in the deck of climate change. There are large differences in the effects of clouds on the radiation balance, depending on the locations of the clouds. Clouds reflect solar radiation back into space and also emit long-wave radiation to space. For low clouds, the reflected solar radiation dominates, so that an increase in low-altitude clouds would lead to an increase in the 77 W/m² of reflected solar radiation, which would have a cooling effect, amounting to a negative feedback and a reduction of global warming. High clouds, on the other hand, are much colder than the underlying surface, and so the long-wave radiation from the clouds is relatively small. An increase in high clouds would lead to a decrease in the 30 W/m² of radiation emitted to space, and therefore would enhance warming (a positive feedback). I also point out that an increase in the water vapor content of the atmosphere need not necessarily lead to increased cloudiness.

What is the bottom line? Greenhouse gases warm the atmosphere and the Earth's surface. This is simple physics. We know that it is true. Positive feedback will make the warming effect stronger. Negative feedback will make it less strong. We simply do not understand some of the feedback very well. Paramount among these is the cloud feedback. But one thing is certain: increasing concentrations of greenhouse gases in the atmosphere has already led to global warming, and it will lead to further global warming in the future.

3.3   CLIMATE MODELS: PREDICTING GLOBAL WARMING

How do scientists determine how the climate will respond to increases in greenhouse gases? In most areas of science, we rely on what we call testable hypotheses. If we want to know the drag force on a falling sphere, we hypothesize that the drag force depends on something like the square of the speed of the sphere and then we go to the laboratory and measure this drag force to test our original hypothesis. If the experimental results agree with the hypothesis, we feel that our conceptual "model" was correct. We can then write down the mathematical equation that Isaac Newton developed, that the gravitational force (the weight of the sphere) minus the drag force is equal to the mass of the sphere times its acceleration, and predict the downward velocity and acceleration of sphere if it is dropped. But how do we test our hypotheses about the effect of greenhouse gases on climate? We have some data on ancient climates, when the atmospheric CO_2 content was much greater or much less than it is today. During the peak of the last ice age, it was considerably colder than it is now, and there was substantially less CO_2 in the atmosphere then. Millions of years ago, there was a very warm period called the mid-Cretaceous, and we know that the atmosphere contained several times the amount of CO_2 that is in today's atmosphere.[4] By comparing these two extreme past climates, Hoffert and Covey estimated that a doubling of CO_2 would eventually lead to a rise in the Earth's temperature of about 4°F.[5] This is about as close as we can come to testing climate model predictions of global warming against the real climate by using ancient data. But we know the essential physics, and, as we shall see, climate models do faithfully describe the changes in climate that have occurred in the 20th century. We also know that greenhouse gases such as CO_2 and water vapor strongly absorb relatively low temperature radiation and that they absorb very little radiation typical of that coming from the Sun. Of this, scientists are absolutely certain. Engineers have been using this fact for more than a century in the design of coal- and gas-fired boilers in power plants.

What is a climate model? Scientists have known for many years how to write the mathematical equations that govern the motions of the atmosphere and the oceans. But we also need to know the "boundary conditions," a set of equations that govern the exchange of latent and sensible heat among the atmosphere, oceans, and land areas, and we do not know how to predict these nearly as accurately. Also, predicting the formation and motion of clouds (which we have seen accounts for a potentially important feedback) is difficult and uncertain. Even so, the equations describing climate are so difficult that they must be solved by programming them on very large and very fast computers. The computers solve the equations by breaking the atmosphere and the ocean into large chunks, frequently many miles or more on a side and a single to several miles deep. As a result, it is difficult for climate models to do a good job of predicting local climate.[1]

One important test for these large complex climate models (called general circulation models or global climate models, in either case, GCMs) is whether they can "predict" the current climate accurately. Alas, although they do a remarkably good job in matching the global climate and that over large areas, such as the size of several states, they do not do a very impressive job locally. This is to say, they predict the current climate pretty well for large regions of the North American continent, but they may not get it right in detail for a particular state. However, as we learn more about how to better model those "boundary conditions" and the formation and motion of clouds and as bigger and faster computers become available, the models are getting better and better. In fact, a decade ago, the horizontal size of the "chunks" was about 300 mi, and the vertical size models have horizontal lengths of about 50 mi, and both the atmosphere and the ocean are separated into some 30 layers.[1]

To see how well the models do on a large scale, see Figure 3.3. The top two figures show model predictions of the zonally averaged surface air temperature for northern hemisphere summer and winter by six climate models. The crosses and open circles represent observational data from two sources. Zonally averaged means averaged over all latitude bands; thus, it means the average value for each longitude. The bottom two figures show predicted precipitation, also zonally averaged, for the same six models compared with observed values. The models are seen to do a fairly good job on these space scales.

To see how well the models perform in predicting globally averaged temperature, see Figure 3.4. The bottom figure shows (in blue) model-predicted temperatures from 19 simulation runs using five naturally forced models (including volcanic eruptions and solar changes). The average of these is shown in the solid dark blue line. Actual globally averaged temperature is shown as the black line. The top figure is produced by 58 simulations using 14 different models with both natural and anthropogenic (man-made) forcing. The yellow lines are model predictions, and the red line is the average of these predictions. Again, the black line represents the observed surface temperature. Note that these values are not the actual temperatures but the departure from the average temperature during 1910 to 1950.

Perhaps, the most difficult thing for climate models to predict is precipitation. Figure 3.5 shows global precipitation as observed and predicted by the Climate Prediction Center merged analysis, which comes from a set of several climate models.

What the scientists who work with GCMs usually report in the scientific literature are the changes in climate that will occur when CO_2 is doubled or quadrupled. This may sound like a more troublesome exercise than it really is. All of the models predict some increase in the globally averaged temperature as well as in the hydrologic cycle (more precipitation and evaporation globally). However, because evaporation as well as precipitation increases and the distribution of both changes, some places end up wetter and some places end up dryer. But the models differ slightly on the details of just where things become wetter or dryer. But one feature of the GCM predictions is common to virtually all the models. They all predict that summer soil moisture will decrease on most continents in regions where it is important to the practice of agriculture.

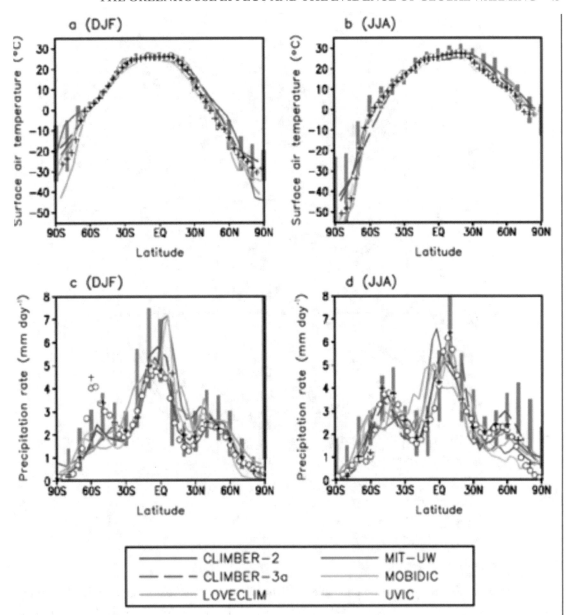

FIGURE 3.3: Latitudinal distributions of zonally averaged surface air temperature (a,b) and precipitation rate (c,d) for northern hemisphere winter (DJF) and summer (JJA) by six relatively simple climate models with atmospheric CO_2 concentrations set at 280 ppm. Observations by two separate groups of scientists are shown as open circles and crosses. The vertical gray lines are results from several more complex GCMs. The square shows the climate models used.

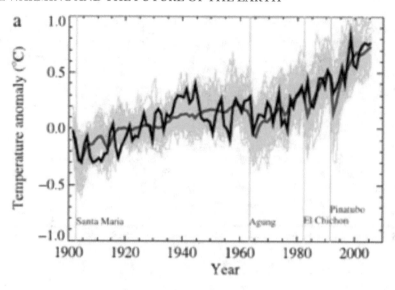

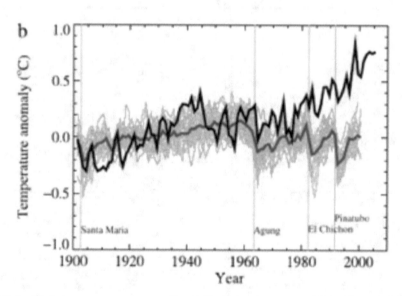

FIGURE 3.4: Global temperature anomalies (°C) from observations (black) and GCM simulations forced by (a) both anthropogenic and natural forcings and (b) natural forcings only. Data shown are global mean anomalies relative to the period 1901–1950. The yellow lines are results from 58 simulations using 14 climate models. The red line is their average. The blue lines are results of 19 simulations from five models, and the dark blue line is their average. In both panels, the black lines are observations.

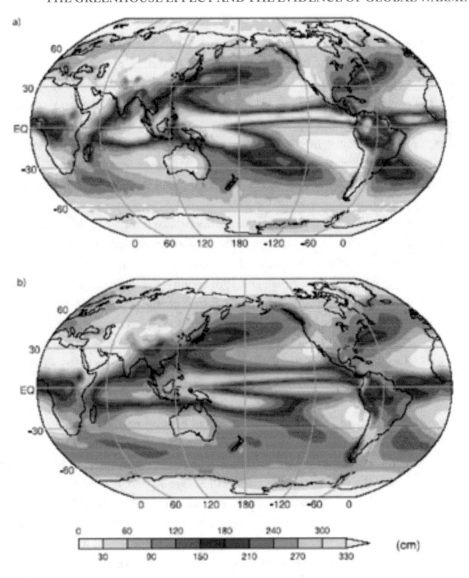

FIGURE 3.5: Annual averaged precipitation in centimeters per year, observed (a) and model predicted (b) based on a multimodel mean from the Climate Prediction Center merged analysis of precipitation. Observations were not available in the small gray regions.

3.4   THE OCEAN DELAYS THE WARMING

Of course, the temperature of the Earth does not immediately respond to increased heating. The ocean has a very large heat capacity; that is, it takes a great deal of heat to warm the ocean only a little. But if the heating rate remains constant, the ocean temperature eventually responds, just as a pot of water does when you heat it on your stove. The climate responds even more slowly if the heat is applied gradually, just as the pot of water does if you turn up the heat on your stove gradually. Because the CO_2 in the atmosphere is building up gradually, global warming will occur gradually. The warming resulting from greenhouse gas increases will occur decades later, and it is certainly not expected to be evenly distributed around the Earth. This is why the temperature changes in Figure 3.4 respond slowly to greenhouse forcing.

Measurements of the temperature of the Earth's surface show a general warming trend during the last 100 or so years, when measured local temperatures were available at enough places that a reasonable estimate of globally averaged temperatures could be computed. There is general agreement among scientists that since the late 19th century, the global temperature has increased by somewhere between four and eight tenths of a degree centigrade (about one half to more than one degree Fahrenheit). The temperature changes during the past century shown in Figure 3.4 rise and fall from year to year and from decade to decade, but the general warming trend is evident. There was a fairly rapid increase in the global temperature between about 1910 and 1940, followed by a leveling off (or even a slight decrease in the northern hemisphere) until the mid-1970s, and a resumption of temperature rise thereafter.

The difference between the responses of land and ocean areas can be seen in Figure 3.6. The land temperature generally responds faster than the ocean surface temperature.

More recently, a group of paleoclimatologists (scientists who study ancient climates) used proxy data to see how the temperatures of the northern hemisphere have changed during the past 1000 years.[6] The definition of the word *proxy* is "one who acts for another." Of course, we cannot directly measure the temperature of the Earth hundreds or thousands of years ago. It happens, however, that there are things that we can measure today that are closely related to what the temperature was long ago. Using proxy data is common among climatologists, especially those who study ancient climates.[4] This particular group examines things such as tree rings in very old wood. Each year, as a tree grows, it makes a ring of new wood that is visually distinguishable from that of the previous year. By measuring the width and the density of each ring and counting back from the year of the most recent ring, scientists can estimate the temperature and the amount of precipitation during the growing season. The width and density of a given ring depend on both temperature and precipitation, but the relative amounts of hydrogen and deuterium seem to depend only on temperature. This allows scientists to separate the two effects. There is, of course, considerable uncertainty in getting the precise temperature. But if there is more than one proxy available, one can do "multiproxy"

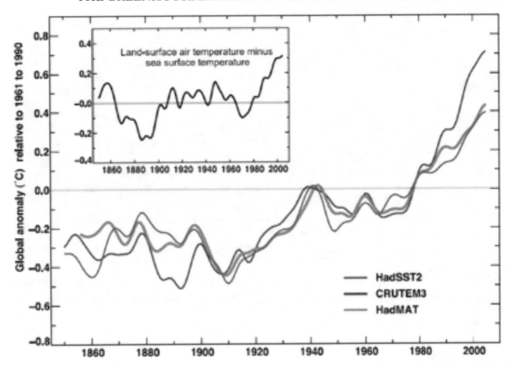

FIGURE 3.6: Variations of the Earth's surface temperature relative to the average of the 1961–1990 period for the last century and a half. These are smoothed data from three scientific groups. The blue line represents globally averaged sea surface temperature; red line, land surface temperature; and green line, combined global surface temperature.

analysis, and the estimates of temperature become more believable the more proxies we have. Other proxies are the growth and chemical analysis of corals, the abundance of sea creatures in particular locations in the sea (depending on what the sea temperature was), the abundance of certain chemical isotopes in the ocean sediments and glaciers, and the extent (growth or melting) of glaciers and ice sheets. The paleoclimatologists from the University of Massachusetts and the University of Arizona used multiproxy analysis to estimate the temperature of the northern hemisphere for the past thousand years. The results were surprising. Their results are shown in Figure 3.7. The farther back in time we go, the more uncertainty we see.

Even considering the uncertainty, we can see two things pretty well. Temperatures drifted slowly downward (cooled) for 900 years until the beginning of the 20th century. Afterward, they began to rise very rapidly. The good news is that we may have been slowly drifting into the next ice age cycle, just as we were predicted to do in accordance with the Milankovitch theory. The bad news is that as we pump more and more CO_2 into the atmosphere, the increase is not likely to stop. The last 20 years or so have been warmer than any such period for the past 1000 years!

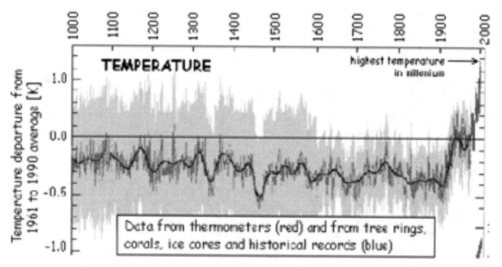

FIGURE 3.7: Northern hemispheric temperature reconstruction during the last 1000 years from proxy data in combination with the instrumental data record.[6] Note the gray area showing great uncertainty.

3.5   NATURAL VARIABILITY

Perhaps, a more detailed discussion of what is meant by natural variability is called for. Both historical data and the results of GCMs show that both the globally averaged temperatures and regional and local temperatures will have their ups and downs. Historical changes that have occurred both recently and during the past thousand years bear out the fact that sizeable changes in the climate might occur naturally in the climate system. This is what climatologists call *natural variability*. For example, there was a relatively cool period that occurred between about 4500 and 2500 years ago, ending at about the time of the dawn of the Roman Empire.[4] The decline and fall of the empire and the beginning of the Dark Ages about A.D. 500 to 1000 saw a return to colder climates. After this period was a relatively warm period called the Medieval Optimum, which lasted from about A.D. 1100 to 1300 in Europe but apparently not in the northern hemisphere as a whole as can easily be seen in Figure 3.7, which shows a rather persistent cooling of the northern hemisphere throughout the period. It most likely occurred as a result of warming of the North Atlantic Ocean. During this period, wine grapes flourished in England, and the Arctic ice pack retreated, to the extent that Iceland and Greenland were settled by Europeans for the first time. About 1450, there was a return to a cooler climate. This period, which lasted through most of the 19th century, is known as the Little Ice Age. One can easily see why some people believe that the recent warming that has occurred is simply the result of natural climate variability as the Earth emerges from the most recent cold period. One of my graduate students recently decided to study past climate variability to see whether he could learn anything about future natural climate trends based on what has happened in the last 1000 years. Based on statistical analysis, we concluded that natural variability should have caused warming of a few tenths of a degree until about 1940, but since that time, the Earth should be cooling. Extrapolating statistical data into the future is risky business at best, but the result provides food for thought because it might refute one of the arguments against the supposition that the recent warming only reflects natural variability. The argument goes like this. During the early 20th century when global emissions of CO_2 were small, the global temperature was increasing rapidly, but between 1940 and 1970, when CO_2 releases had grown substantially, the increase in global temperature was nil. If this preliminary hypothesis is correct, the increase in temperature early in the century was caused mostly by natural variations in the climate. Later on, the natural tendency for the climate to cool was opposed to the greenhouse effect, and the climate neither warmed nor cooled. After about 1970, the climate began to warm again, as the greenhouse warming became stronger than the natural cooling. But why did the cooling episode occur? I believe that it is variability of the thermohaline circulation. It happens that the ocean temperatures at depths of 500 to 2000 m in both the North Atlantic and North Pacific warmed while the surface temperatures were cooling during this period. The deep ocean was taking all the heat! I will say more when in the next chapter, I respond to the skeptics' Challenge 12. In any case, detecting a greenhouse warming signal within the historical temperature records is clear. It is clear that the globally averaged temperature of Earth's surface has risen during the present century, and most importantly, it has done so over a shorter time period than at any time since the end of the last ice age about 15,000 years ago.

The warming of about eight tenths of a degree centigrade that has occurred during the last hundred years is faithfully predicted by the GCMs. It is certain that the climate of the Earth will warm as the atmospheric concentrations of greenhouse gases increase. It is very likely that it will warm, on average, at about the rate predicted by the GCMs that exhibit medium sensitivity. These models predict that the global temperature will increase by 2°C to 5°C by 2100, depending on how committed we are to reducing greenhouse concentrations. (I will be more specific about this later when I discuss various energy use scenarios.) Moreover, it is important to remember that high latitudes will warm as much as two to three times as much as this globally averaged value. This we know from both models and historical data.

3.6   FINGERPRINTS: OBSERVATIONS OF GLOBAL CLIMATE CHANGE

Let us now examine some of the observational evidence of global warming so far. What should we have seen? Some evidence of what to expect comes from GCM predictions but also comes from what we know about ancient and fairly recent climates, for example, the recent warm and cool periods described above as well as the most recent ice age from about 120,000 to 15,000 years ago.

One may perhaps be somewhat suspicious of claiming with confidence that changes in a single variable demonstrate a cause-and-effect relationship. If the climate of the Earth is indeed warming because of increased concentrations of greenhouse gases, the evidence should show up in other ways besides the globally averaged temperature near the surface. Examining many such signals is often referred to as the "fingerprint method." For example, the growing season should be getting longer, and the amount of snow that falls in midlatitudes in the spring and fall should be decreasing because winter weather comes later and spring earlier. Examining all of these things that scientists predict will happen gives us an even better way to see whether the global greenhouse effect is beginning to take hold. It turns out that nearly all of the things that we think ought to be happening (what our models and date from other warm and cool periods and our intuition suggest) are happening.

3.7   HERE IS WHAT WE SHOULD EXPECT

- Obviously, the surface of Earth and the atmosphere should warm. The ocean surface should warm more slowly than land areas because the ocean has a larger capacity to absorb some of the heat.
- The stratosphere should cool if the greenhouse effect is responsible for the warming, and it should warm if solar variations are the culprit. This is shown by simple radiation models and is very well understood.
- High latitudes should warm more than lower latitudes. Models show this. More important, evidence from past warmer climates shows this to be true.
- Simple physics indicates that warmer air can hold more moisture, so the atmospheric water vapor content should increase.
- This leads to more precipitation. However, the increase in precipitation will not occur everywhere, and because of increased evaporation brought on by the higher temperatures, some land areas will become dryer. Thus, more intense and longer droughts are predicted in some areas, particularly areas where it is already dry.
- Rainfall will tend to occur more often in the form of very intense events.
- In response to increasing temperatures, the growing season should become longer in mid to high latitudes. Migration patterns of birds and animals will be affected.
- Permafrost in regions where it exists should decrease in extent.
- Mountain glaciers in mid and high latitudes should retreat. Eventually, ice caps in Greenland and Antarctica should begin to melt.
- Sea ice should retreat in both hemispheres.
- Snowfall should decrease in all but possibly very high latitudes. At very high latitudes it will be cold enough for snowfall, at least for a while during winter, and the increased atmo-

spheric water vapor might even cause increased snowfall. However, the poleward extent of snow should decrease in both hemispheres.

Here are some of the things that have been observed. (You may go directly to Table 3.1 to see a compilation of what we have observed.)

- Eleven of the last 12 years (1995–2006) rank among the warmest years in the instrument record of the global surface temperature since 1850. The total globally averaged temperature increase from 1850–1899 to 2001–2005 is approximately 0.76°C. The linear warming trend during the last 50 years (0.13°C per decade) is nearly twice that of the last 100 years. Urban heat island effects are real, but local, and have a negligible effect (<0.006°C per decade over land and zero over oceans) on these values. (See Challenge 8 in Chapter 4.)
- New analyses of balloon-borne and satellite measurements of lower tropospheric and midtropospheric temperature show warming rates that are similar to those of the surface temperature record and are consistent within their respective uncertainties, reconciling discrepancies noted by Spencer and Christy. (See Challenge 7 in Chapter 4.)
- The averaged water vapor content has increased at least since the 1980s over land and ocean as well as in the upper troposphere. The increase is broadly consistent with the extra water vapor that warmer air can hold.
- Observations since 1961 show that the average temperature of the global ocean has increased to depths of at least 3000 m and that the ocean has been absorbing more than 80% of the heat added to the climate system. Such warming causes sea water to expand, contributing to sea level rise. We now know that glaciers have been retreating during the past century. The overall area of glaciers in the Alps is about half what it was in the 1850s. Additionally, tropical glaciers in the Andes, on the North American continent, and in Asia have been retreating. Mountain glaciers and snow cover have declined on average in both hemispheres. Widespread decreases in glaciers and ice caps have contributed to sea level rise. In the past, few decades some of the great Himalayan glaciers have either disappeared or have eroded considerably. In the Saraswati valley, north of Badrinath, the Ratakona glacier, which lies near the Mana Pass, is on the verge of disappearing. Similarly, the Pindari and Milan glaciers are also gradually receding. Figure 3.8 illustrates the evolution over time of the Gangotri glacier, which is the source of the Ganges river. Since 1956, it has been receding at such an alarming rate that some environmentalists fear that a time may come when the Ganges itself disappears. For the last half century, it has been receding at a rate of about 30 m/yr. During the last decade of the 20th century, the glacial channel feeding the Ganges River shifted 20 m, and the volume of water is shrinking rapidly.

Figures 3.9–3.11 show "before-and-after" photographs of two Alaska glaciers and one from Peru. Particularly striking is the Muir glacier, which has retreated so far since 1941 that one has to look carefully to see its tip in the background of the right photograph. The Qori Kalis glacier in Peru has also receded and formed a glacial lake.

The Triftgletscher in Switzerland is shown in Figure 3.12. The rapid melting occurred in only 1 year!

In the 1850s, nearly 4500 km² of the Alps were glaciated. By the 1970s, the area had fallen to under 3000 km², a loss of about 3% per decade. From the 1970s to 2000, the rate of loss had increased to more than 8% per decade.

TABLE 3.1: Observations of global warming
Globally averaged temperatures are increasing at a rate of 0.13°C per decade. Eleven of the last 12 years were the warmest since temperatures have been recorded.
Satellite measurements show that the troposphere and the surface are warming simultaneously. The lower stratosphere has cooled, ruling out solar increases as the cause.
The water vapor content of the atmosphere has increased
Mountain glaciers are retreating dramatically almost everywhere. Several glaciers in Greenland and Antarctica are flowing to the sea at increasing rates.
Temperatures at the top of the permafrost in high latitudes have increased, and the area of seasonally frozen ground has decreased.
The area of the Arctic sea ice has decreased dramatically as has its thickness.
Birds, mammals, fish, and plants have responded by migrating, flowering, and fruiting earlier, indicating earlier spring and later autumn.
The date of the last snowfall is earlier. The area of maximum snow cover in the northern hemisphere has decreased by 10% in the last century.
Widespread changes in extreme temperatures have been observed. Heat waves have become more frequent.
Tropical storms have increased both in number and in intensity.
Extensive bleaching of corals is occurring as a result of increasing ocean temperatures.

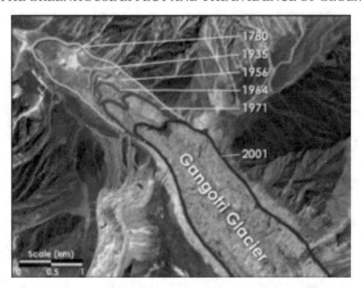

FIGURE 3.8: Time evolution of the Gangotri glacier in the Himalayas (source NASA)

FIGURE 3.9: McCall glacier, Alaska: 1958–2003: (Byrd Polar Research Center, Ohio State University).

FIGURE 3.10: Muir glacier, Alaska: 1941–2003 (Byrd Polar Research Center, Ohio State University).

- Many of the glaciers in Glacier National Park are in grave danger of disappearing as did the Grinnell glacier shown in Figure 3.13. You can find many more of these glacier pairs by going to *Glaciers Online* (http://glaciers.research.pdx.edu/). Figure 3.14 illustrates the recession of continental glaciers worldwide. Note that European glaciers as a whole have not receded, although those in the Alps have. Some Scandinavian glaciers have advanced.
- Temperatures at the top of the permafrost layer have generally increased since the 1980s in the Arctic (by up to 3°C). The maximum area covered by seasonally frozen ground, shown

FIGURE 3.11: Qori Kalis glacier, Peru: (Byrd Polar Research Center, Ohio State University).

FIGURE 3.12: Rapid recession of Triftgletscher: top, 2002; bottom, 2003 (source Glaciers Online).

in Figure 3.14, has decreased by 7% in the northern hemisphere since 1900, with a decrease in the spring of up to 15%.[1]

- New data now show that losses from the ice sheets of Greenland and Antarctica have very likely contributed to sea level rise during 1993–2006. Flow speed has increased for some Greenland and Antarctic outlet glaciers, which drain ice from the interior of the ice sheets. Satellite measurements show that the amount of ice melt flowing into the sea from large glaciers in southern Greenland has more than doubled in the last 10 years. Glaciers normally flow very slowly. The rapid increase in the flow may come about when the bottom of the ice is lubricated by melt water cascading downward through crevasses in the ices called moulins, like the one shown in Figure 3.15. Climate models currently do a poor job of predicting ice sheets because they do not take this into account a new study in the journal *Science* that suggests that sea levels will rise by 20 to 55 inches by 2100.

FIGURE 3.13: The Grinnell glacier in Glacier National Park.

The most recent data on ice loss from Greenland is shown in Figure 3.16. This figure was presented in a talk at the American Society of Mechanical Engineers last year by Doctor Michael Schlesinger. Global average sea level rose by an average of 1.28 mm/yr during 1961–2003. The rate was faster during 1993–2003, about 3.1 mm/yr. There is high confidence that the observed rate of sea level rise increased from the 19th to the 20th century. The total 20th century rise is 0.17 m. The recent fourth report of the IPCC reported that by 2100 sea level would rise between 5 and 23 cm, or 2 to 10 inches. However, the calculations implicitly assume that none or very little of the increase will come from melting of the Greenland and Antarctic sheets. At present, Greenland is losing about 57 mi^3/yr of ice, capable of raising sea level by 0.6 mm/yr.

Average Arctic temperatures increased at almost twice the global average rate in the past 100 years. In many places, it is much larger, approaching 3°C over northern Greenland. Satellite data since 1978 show the annual average Arctic sea ice extent has shrunk by 2.7% per decade, with larger decreases in summer (7.4% per decade). Both the extent and thickness of northern hemisphere sea ice have decreased recently based on satellite measurements and that of southern hemisphere sea ice has probably decreased also, although this is not quite as certain. The decrease in the extent of northern hemisphere sea ice is based on satellite measurements, and the decrease in

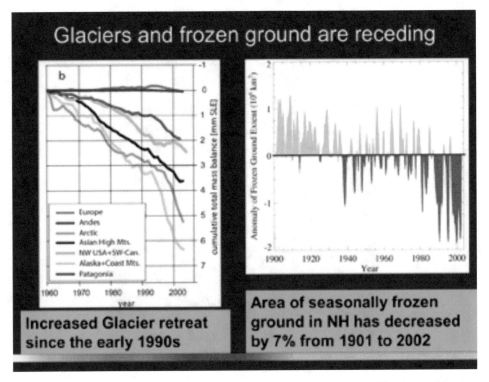

FIGURE 3.14: Glacier retreat for continental glaciers (left) and decrease in seasonally frozen ground (right).

thickness is based on measurements from submarines. In the 1990s, the U.S. Navy allowed scientists to go on five data-taking cruises under the Arctic sea ice. Data from such submarines have been classified in the past,[9] but recently, some data became publicly available from cruises that took place between 1958 and 1976. When the two data sets were corrected for seasonal growth and compared, it was discovered that the average ice thickness had decreased by 1.3 m. The decrease in the extent of southern hemispheric sea ice is based on a rather ingenious "proxy" described by de la Mare.[10] It happens that mink whales keep very close to the edge of the sea ice in the southern hemisphere. By examining the records of whaling ships in the early part of the last (20th) century, these scientists were able to find where the sea ice edge was located before the existence of the satellite data. The declining area of Arctic sea ice is illustrated in Figure 3.17, and satellite pictures are shown in Figure 3.18. The latest data presented at the American Geophysical Union Fall 2007 Meeting in Vienna suggests that the ice is no longer showing a robust recovery from the summer melt. Last month, the sea that was frozen covered an area that was 2 million km² less than the historical average, an area larger than the size of Alaska. The sea ice reached its minimum extent on September 14, 2006, making 2006 the lowest on record in 29 years of satellite record-keeping. Most climate

FIGURE 3.15: A moulin in Greenland.

models' simulations have predicted a loss in September ice cover of 2.5% per decade from 1953 to 2006. The fastest rate of September retreat in any individual model was 5.4% per decade. (September marks the yearly minimum of sea ice in the Arctic.) But newly available data sets, blending early aircraft and ship reports with more recent satellite measurements that are considered more reliable than the earlier records, show that the September ice actually declined at a rate of about 7.8% per decade during the 1953–2006 period. This suggests that current model projections may, in fact, provide a conservative estimate of future Arctic change and that the summer Arctic sea ice may disappear considerably earlier than IPCC projections. The 2006 IPCC reports that Arctic sea ice may disappear by sometime between 2050 and 2100. It may disappear much earlier.

- In 2002, an ice shelf known as Larsen B disintegrated in just over a month. Ice shelves are thick plates of ice that reach into the sea and are fed by glaciers. The Larson B sheet was about 200 m thick, and the ice that was released in this short time weighed about 720 billion tons. The collapse is attributed to the strong warming, about 0.5°C per decade, a trend that has been present since the 1940s. It appears to have been related to the presence of melt water ponds on top of the ice shelf during late summer. In addition to absorbing more

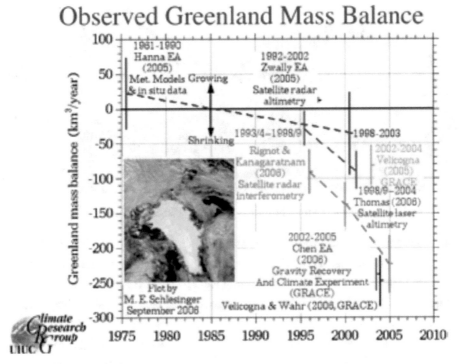

FIGURE 3.16: Recent data on Greenland ice loss.

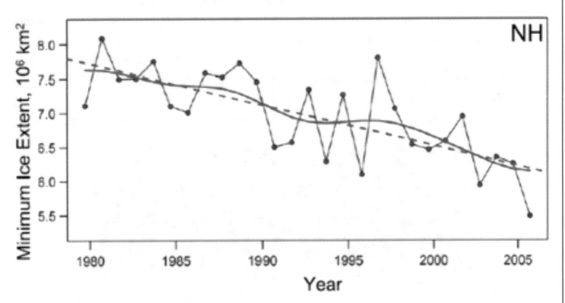

FIGURE 3.17: The decrease in the extent of Arctic sea ice in summer since 1970. The blue curve shows smoothed decadal variations, and the dashed line is the linear trend of –7.4% per decade.

Sea Ice Minimum 1979   Sea Ice Minimum 2005

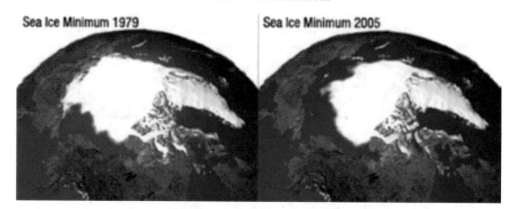

FIGURE 3.18: Satellite views of annual minima in Arctic sea ice, 1979 and 2005.

solar heat than does ice, the water probably enhanced fracturing of the ice by filling cracks in the ice and forcing the heavier water through to the bottom. Melting ice shelves do not increase sea level because they are floating ice. However, it serves to demonstrate how very rapidly and unexpectedly things can change in ice. Figure 3.19 shows the collapse as captured by NASA's Terra satellite.

- Coral reefs are bleaching. Coral reefs get bleached when the water gets too warm. The two largest coral reefs are the Belize Barrier Reef (largest in the western hemisphere) and the Australian Great Barrier Reef. The Belize reef has suffered 40% damage since 1998, not because the average temperature of the water has increased very much, but because of short intervals of high water temperatures. The Great Barrier Reef suffered extensive bleaching in 2002, when the water temperature was higher than usual, and again in 2006. Scientists estimate that a quarter of the world's coral has been permanently lost, with more lost every year.

- Scientists have observed a wide range of the migration patterns of birds, fish, and turtles, apparently in response to warming. Observations of plants, animals, and birds in the Arctic

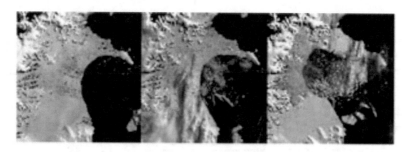

Larson B Jan 31   Larson B Feb 23   Larson B March 5

FIGURE 3.19: The collapse of the Larson B ice shelf in 2002.

revealed that flowering and egg-laying occurs some 2 weeks earlier in response to the early emergence of spring. Ice in northeast Greenland is melting 14 days earlier than it did in the mid-1990s. There are many studies detailing the early onset of seasonal events due to global warming. For example, in Scotland, butterflies arrive 8 days earlier than they did 30 years ago. Birds are nesting as much as 10 days earlier. A recent study by scientists from 17 nations examined changes in recurring natural events such as when plants flowered. The results were published in the journal *Global Change Biology* and showed that there was a direct link between rising temperatures and changes in plant and animal behavior. The scientists examined 125 000 studies involving 561 species in European countries from 1971 to 2000. The study showed that 78% of all leafing, flowering, and fruiting were happening earlier in the year.

- Changes in precipitation and evaporation over the oceans are suggested by freshening (becoming less salty) of mid- and high-latitude waters together with increased salinity in low-latitude waters.

- Climate models and some data from warmer climates indicate that some regions will become wetter, whereas, despite the increased atmospheric water, some regions will become dryer because of increased evaporation from the warmer surface. Significantly increased precipitation has been observed in eastern parts of North and South America, northern Europe, and northern and central Asia. Drying has been observed in the Sahel, the Mediterranean, southern Africa, parts of southern Asia, and Australia. In agreement with climate model predictions, more intense and longer droughts have been observed over wide areas since the 1970s, particularly in the tropics and subtropics. The frequency of heavy precipitation events has increased over most land areas, consistent with warming and increases in atmospheric water vapor. Figure 3.20 shows conditions in the United States. Note that the region around Texas and Oklahoma is currently experiencing extreme to severe drought, whereas, during June and July 2007, the region suffered severe flooding. In China, there are similarities. At least 18 million people have been affected by the worst drought in 50 years. The southwestern region of Chongqing has been worst hit, but areas of Sichuan and Liaoning are also affected. In Chongqing, there has been no rain in several months, and two thirds of the rivers have dried up. Yet, earlier this year, parts of China were hit by heavy rainfall and flooding. Despite this, the increased evaporation has actually reduced water availability. Wetlands in the Qinghai–Tibet Plateau have shrunk by more than 10%, with those at the origin of the Yangtze River suffering most. Water flow in the Yangtze River is down by 40%, and the two largest lakes in the region, Dongting Lake and Poyang Lake are 60% and 10%, respectively, lower than their normal levels. More than 2000 lakes in China have dried up. On the other hand, India is suffering from flooding

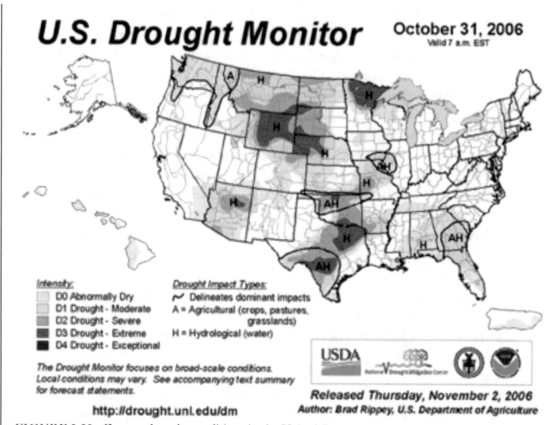

FIGURE 3.20: Current drought conditions in the United States.

from glacial lake outburst floods, which are catastrophic discharges of water due to melting of the glaciers in the Himalayas, although large parts of the country are experiencing droughts. Further bad news is that once most of the glaciers are gone, water availability will be even more scarce. A dry spell in Morocco has slashed the country's 2007 grain crop to an estimated 2.0 million tons from 9.3 million last year, and the government is expected to triple soft wheat imports to 3.0 million tons. Droughts and floods are already becoming more frequent, and projections are they will become more so in the future. Closer to home, the long-term drought in southern Florida has caused the water level in Lake Okeechobee, a vital source of water (as well as wildlife), to shrink to a new low.

- Snowfall in a limited area of East Antarctica has increased recently, as it should under global warming because the poleward transport of moisture is predicted to increase. However, the end of the snow season is occurring earlier in the northern hemisphere, as shown in Figure 3.21. Although snowfall can increase under warmer conditions at very high lati-

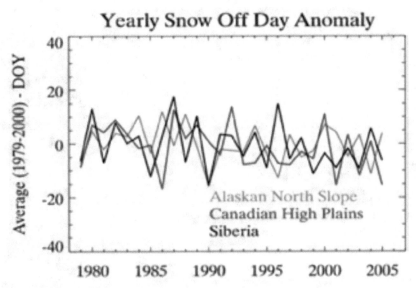

FIGURE 3.21: The date of the last snowfall of the year is earlier.

tudes, precipitation at somewhat lower latitudes can change from snow to rain if the local temperature increases, resulting in a decrease of snow cover. The extent (area) of snow cover in the northern hemisphere has decreased by about 10% over the last century.[1]

• Widespread changes in extreme temperatures have been observed during the past 50 years. Cold days, cold nights, and frost have become less frequent, whereas hot days, hot nights, and heat waves have become more frequent.

• There is observational evidence for an increase of intense tropical cyclone activity in the North Atlantic since 1970. There is increasing evidence that both the number of tropical storms and the number of very intense ones have increased. This makes sense because tropical storms are driven by evaporation from the ocean, and the tropical ocean has become warmer. Trends since the 1970s show an increase in storm duration and greater storm intensity. The number of category 4 and 5 storms has increased by about 75% since 1970. Until the hurricane season of 2006, which was rather mild, the numbers of hurricanes in the North Atlantic were above normal in 9 of the past 11 years (Figure 3.22).

The climate change fingerprint is there. When all of the evidence is taken together, it leaves no doubt that Earth's climate has warmed during the last century.

The bottom line? The globally averaged temperature of the Earth has increased (although not continuously) by about half a degree centigrade during the last 100 years. This is entirely consistent with estimates from the reconstruction of climates when the CO_2 content was different from the

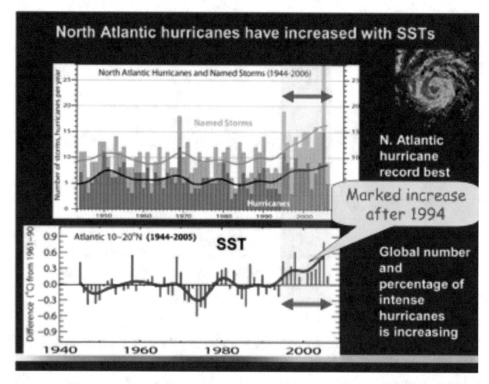

FIGURE 3.22: Atlantic hurricanes have increased both in number and intensity (source: Matthew Huber, *Geophysical Research Letters*, June 1, 2006).

present and with the results of GCMs. One statistical study implies that the natural variability should quite possibly be cooling at present, whereas the presence of sulfate aerosols should almost certainly be causing a cooling, especially in the northern hemisphere. Yet, the long-term trend shows a definite warming. Other data provide an identifiable fingerprint of global warming. It is certain that the warming will continue as more greenhouse gases enter the atmosphere.

NOTES AND REFERENCES

Figures 3.1, 3.3–3.7, 14, 17, 18, 21, and 22 are from Reference 1. Figures 3.15 and 3.16 are from Michael Schlesinger, University of Illinois at Urbana–Champaign.

1. Forster, P., Ramaswamy, V., Artaxo, P., Berntsen, T., Betts, R., Fahey, D.W., Haywood, J., Lean, J., Lowe, D.C., Myhre, G., Nganga, J., Prinn, R., Raga, G., Schulz, M., and Van Dorland, R., "2007: Changes in atmospheric constituents and in radiative forcing," in *Climate Change 2007: The Physical Science Basis. Contribution of Working Group I to the Fourth Assessment*

Report of the Intergovernmental Panel on Climate Change, S. Solomon, D. Qin, M. Manning, Z. Chen, M. Marquis, K.B. Averyt, M. Tignor, and H.L. Miller, eds., Cambridge University Press, Cambridge, UK, p. 130, 2007.

2. See Watts, R.G., "Climate models and the prediction of CO_2-induced climate change, *Climatic Change*, vol. 2, 1980, pp. 387–408, for a full discussion and an estimate of various climate feedbacks.

3. Modest, M.F., *Radiative Heat Transfer*, McGraw-Hill, New York, 1993.

4. Crowley, T.J., and North, G.R., *Paleoclimatology*, Oxford University Press, Oxford, U.K., 1991.

5. Hoffert, M.I, and Covey, C., "Deriving global climate sensitivity from paleoclimate reconstructions," *Nature*, vol. 360, pp. 573–576, 1992.

6. Mann, M.E., Bradley, R.S., and Hughes, M.K., "Global scale temperature patterns and climate forcing over the past six centuries," *Geophysical Research Letters*, vol. 26, pp. 759–762, 1999. This has since been updated to include 10 centuries.

7. Mahasenan, N., Watts, R.G., and Dowlatabadi, H., "Low-frequency oscillations in temperature proxy records and implications for recent climate change," *Geophysical Research Letters*, vol. 24, pp. 563–566, 1997.

8. Roemmich, D., and Wunsch, C., "Apparent changes in the climate state of the deep north Atlantic Ocean," *Nature*, vol. 307, pp. 447–450, 1984.

9. Rothrock, D.A., Yu, Y., and Maykut, G.A., "Thinning of the arctic sea-ice cover," *Geophys. Res. Lett.*, 26(23), 3469–72.

10. De la Mare, W.K., "Abrupt mid-twentieth-century decline in Antarctic Sea ice extent from whaling records," *Nature*, vol. 389, pp. 57–60, 1997.

• • • •

CHAPTER 4

The Skeptics: Are Their Doubts Scientifically Valid?

An argument frequently used by members of congress and others for delaying any action to prevent (or slow down) global warming is that the scientific community is divided on the issue of the science of global warming. If scientists cannot agree on such a fundamental question as "Is the greenhouse effect real?," then certainly we should not base policy decisions on the prospects of global warming. The immediate response to this argument is that by any measure the overwhelming majority of climatologists agree on the fundamental science. Policy makers and individuals do and should want to hear both sides of important questions, but they should be aware that although thousands of scientists believe that the greenhouse effect is real, that warming is now occurring, and that the consequences will be very serious and harmful to the environment and to humans, only a very few are now challenging that view. Even so, the answers to scientific questions should not be based on popularity polls. When a serious scientist poses serious questions, these must be dealt with seriously and objectively. Former Vice President Al Gore has stated in his documentary, *An Inconvenient Truth*, that, although hundreds of scientific articles about global warming have been published in recent years, not a single article has appeared in the peer-reviewed literature that is skeptical of global warming or its impacts. This is, in fact, not true. It is true that most of the comments from the skeptics have appeared in the so-called gray literature (not reviewed by scientific peers), but, as we will see, some have appeared in peer-reviewed journals, and these claims must be taken especially seriously. The purpose of this chapter is to list as many of these questions and challenges as I could uncover, those that have been published in peer-reviewed journals as well as those from the gray literature, and to detail the scientific response when a sound one is available. In most cases, the challenges are very easy to refute and require little comment other than some straightforward ones. In other cases, mostly those that have appeared in the refereed scientific literature, a good deal of discussion is required, but the result is the same: each of the claims by the skeptics can be refuted. (I note that it is particularly important to explain and refute the claims made in the gray literature in a language easily understood by nonexperts because that is what the average layperson reads and so forms his opinions.)

Challenge 1. Is the increase in atmospheric CO2 coming from the burning of fossil fuels?
In an October 19, 1997, speech at the World Petroleum Congress in Beijing, Lee R. Raymond, chairman and chief executive officer of the Exxon Corporation, stated that "only 4% of the carbon dioxide entering the atmosphere is due to human activities—96%—comes from nature."[1]

   This is an excellent example of a statement that is true but terribly misleading. We discussed this question in Chapter 2. Figure 2.4 in Chapter 2 puts this question to rest. The *increased* CO_2 in the atmosphere is coming from burning fossil fuels. The decrease in the $^{13}C/^{12}C$ ratio proves once and for all that this is true.

Challenge 2. The atmospheric concentration of CO_2 during the peak of the last glacial cycle, when the Earth was much colder than the present and much lower than it was before the present increase occurred. Did this help to cool the Earth?[2]
Doctor Patrick Michaels states in his book, *Sound and Fury: The Science and Politics of Global Warming*, that "gas bubbles trapped in Antarctic ice tell us that the temperature dropped *before* the CO_2 concentration changed, not after" [emphasis his].[2]

   What Dr. Michaels is unwittingly proposing is that there might be a different link between temperature and CO_2. Did the CO_2 decrease cause the cooling or did the cooling cause the decrease in CO_2? He is insinuating that the cooling might have caused the decrease in atmospheric CO_2 rather than the other way around. We need to go back to the idea of feedback to explore the consequences of just how the two effects connected. In fact, carbon cycle models indicate that on balance carbon is absorbed by cold ocean water and emitted by warm ocean water. If the ocean cooled (as it did during the last glacial maximum), it makes sense that more CO_2 would be absorbed in the oceans at the expense of the atmospheric concentration. If this is true, it implies that as the ocean warms because of increases in atmospheric greenhouse gas concentrations, it is likely to absorb less CO_2, causing atmospheric CO_2 levels to rise even more. This is a scary proposition. A positive feedback not now incorporated in climate models could make global warming even more severe!

Challenge 3. CO_2 concentrations in the atmosphere have been far higher in the past.[2]
Several of those who are skeptical of global warming have made this rather curious statement. Apparently, it is meant to assure us that the Earth survived during these times, and so we can survive larger concentrations of CO_2 in the future. The statement is true but irrelevant, except that we can and have used information about past climates that were warmer because of larger atmospheric greenhouse gas concentrations to get some idea about the future climate. Incidentally, the fact that when CO_2 concentrations were large, the climate was much warmer supports the theory of global warming.

Challenge 4. *There has been no warming in the contiguous United States from 1895 until 1985.*
In a report by the George C. Marshall Institute, *Global Warming: What Does the Science Tell Us?*, Jastrow et al. state that temperature records in the United States reveal no statistically significant warming in the past 100 years, whereas climate models predict a pronounced warming trend there.[3] This they report after spending a great deal of effort pointing out that GCMs, because of their large-grid scales, do not faithfully predict local climate change! Evidently, they must be referring to some local data, or perhaps, they were referring to data collected before about 1980. In fact, as reported in the fourth report of the IPCC, since about 1970, the North American continent has warmed considerably. Figure 4.1 is taken from the IPCC report (see also Challenge 12 for a possible explanation of why the temperature of the region, and the globe, decreased for a time).

Challenge 5. *The temperature record of warming correlates with sunspot number and solar constant variations.*[3]
The authors of the Marshall Institute report suggest that the variations in global temperature in the 20th century are most likely caused by changes in the solar constant. The National Solar Laboratory reported that there are variations in the Sun's radiation as measured at the distance of the Earth from the Sun of about 1.6 W/m². The intensity of solar radiation at this distance from the Sun is about 1360 W/m², so 1.6 W/m² is equivalent to an increase of a little more than 0.1%. In

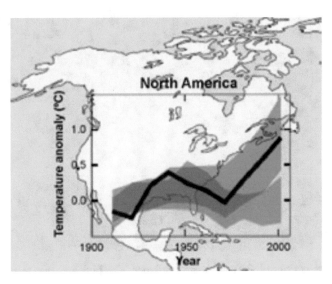

FIGURE 4.1: North America experienced some cooling during about 1940 and 1970. Since that time, it has warmed considerably.

fact, a new reconstruction of solar irradiance based on a model of solar magnetic flux variations by Wang et al. suggests that the variation is significantly less. The reconstruction of total solar variance is shown in Figure 4.2. The most likely increase in solar radiation during the past 300 years is about 0.5 W/m², which corresponds to a radiative forcing of 0.125 W/m². (The value 0.5 has to be divided by 4 to account for the fact that only half the earth is irradiated (i.e., during daylight hours) and also for the Earth's curvature.) GCM results indicate that when models are run until they come to equilibrium, a 2% increase (40 times the reported solar variations) would be required to raise the Earth's temperature as much as doubling greenhouse gas concentrations. Even to cause the recent warming of about half a degree centigrade, an increase in the solar constant of at least 15 times the reported variation would be required. The authors of the Marshall Institute report speculate that there may have been a much larger variation of solar radiation during the last 100 years, but there is absolutely no evidence to support this speculation. In fact, it is seen in Figure 4.2 that no such variation has occurred at least in the past 300 years. Finally, I note that if the warming were because of

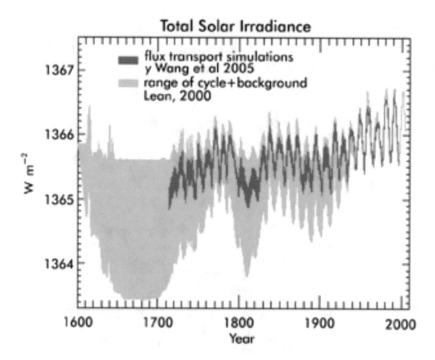

FIGURE 4.2: Reconstructions of the total solar irradiance. The blue area was reconstructed by Lean using brightness changes in Sun-like stars. The red areas are from the more recent reconstruction by Wang et al.[4] based on solar considerations.

increased solar radiation, both the troposphere (the lower atmosphere) and the stratosphere would warm, whereas, if greenhouse gases are responsible, the troposphere would warm and the stratosphere would cool. As you will see when I discuss Challenge 7, the latter is what actually happened.

There is also a slightly different aspect of the claim that the Sun is somehow to blame for global warming. Cosmic rays from other bodies in the universe constantly bombard the Earth. These can create electrically charged ions in the atmosphere that act as magnets for water vapor, and they act as "nuclei" and aid in the formation of clouds. Svensmark, of the Danish Meteorological Institute, has claimed that when the Sun is active (there is an 11-year cycle of sunspot activity), it deflects cosmic rays away from Earth, thereby reducing cloud cover.[5] Yet, there seems to be no correlation between cosmic rays and cloud cover after 1990. Recently, Mike Lockwood and Claus Frohlich brought together solar data for the past 100 years and looked for a correlation between solar variation and global mean temperatures.[6] Solar activity peaked between 1985 and 1987. Since then, trends in solar irradiance, sunspot number, and cosmic ray intensity have all been in the opposite direction to that required to explain global warming. Ken Carslaw, an atmospheric scientist at the University of Leeds in the United Kingdom, has even pointed out that if solar effects are there, they have been responsible for cooling and have been overwhelmed by greenhouse gases.[7] This would mean that future warming might be even greater than is generally thought if the countervailing solar effect comes to an end.

The science behind the cosmic ray hypothesis has for a time been hotly disputed. But this recent work seems to put the nail in the coffin.

Challenge 6. *Three quarters of the full warming should occur within the first 10 years. The time lag induced by the ocean is only a few decades.*[3]

This is also cited in the Marshall Institute report. It is completely without basis. The authors write endlessly about how the large heat capacity of the ocean delays the onset of global warming but apparently completely misunderstand how this works. In the early days of climate modeling, the models were run first without doubled CO_2 and then with doubled CO_2. In other words, the CO_2 was abruptly doubled, and the models run until they reached equilibrium. When run this way, the models reached about three fourths of the equilibrium climate in about 10 years. CO_2 does not suddenly double. In fact, it is increasing rather slowly. It has taken 200 years to increase from about 280 ppmv to its current level of more than 360 ppmv. When the greenhouse forcing is applied slowly, the temperature change predicted by the models lags behind the forcing by many decades, largely because when the change is so slow that heat leaks into the deep ocean more readily than when the forcing is applied suddenly. The transient temperature response of the Earth to currently increasing greenhouse gases simply cannot be understood by examining the early transient experiments in which the CO_2 content was abruptly doubled. The interaction between the atmosphere and the ocean is completely different.

Challenge 7. *In 1979, satellites began measuring the temperature of the lower atmosphere with high accuracy by measuring the vibration frequency of oxygen molecules. Several of the skeptics (e.g., Michaels in Sound and Fury,[2] pp. 31 and 32) state that this record shows no warming since 1979, and is in stark disagreement with ground-based estimates, which show large temperature increases.[2]*

Michaels is referring to an article published in the journal *Science*[8] in 1990 and in several articles after that. The article used satellite measurements to show that the temperature of the lower troposphere (the lower part of the atmosphere) between 1979 and 1990 has increased by only 0.001°C. In a later article, Spencer and Christy claimed that the satellite data showed that the global temperature, from 1979 through 1995, decreased by about 0.05°C per decade.[8] Because these claims were published in major referred journals they cast doubt on the notion that the Earth is warming. If the surface temperature is increasing, the temperature of the lower atmosphere must also be increasing. At the time, this was considered to be a real conundrum. How do we explain these differences?

Measuring temperature is always done indirectly. With the common household thermometer, we measure the length of a column of liquid, colored water, mercury, or some other liquid in a narrow tube. The liquid expands in a predictable way when its temperature increases. The thermometer is calibrated by marking off places in the column to indicate the temperature. This is an easy thing to do. We read the length of the column, and it is directly related to temperature, but what we are reading is the length of the column of liquid, not temperature. What, correspondingly, is the satellite measuring?

Since 1979, a series of nine different polar orbiting satellites have carried something called microwave sounding units (MSUs) that measure the radiation emitted by atmospheric oxygen molecules in a certain frequency range. Beginning in 1998, another group of instruments called advanced MSUs began operation. Changes in the strength of this radiation is related to the temperature of the oxygen molecules and, therefore, to the temperature of the atmosphere itself. A major challenge for climate assessment is merging together data from all the satellites. The merging must account for a number of possible sources or error. These include the following: 1) offsets in calibration between different satellites, 2) orbital decay (satellites gradually fall nearer to the Earth as they slow down), 3) drift in satellite calibrations for different satellites. Several different teams of scientists have used different methods to correct these and other sources of error, and establish a new set of temperature records.[9] The nature of the measurements by the MSUs is such that they do not measure local temperatures, but rather average temperatures in thick layers of the atmosphere. Figure 4.3 shows results from the various science teams.

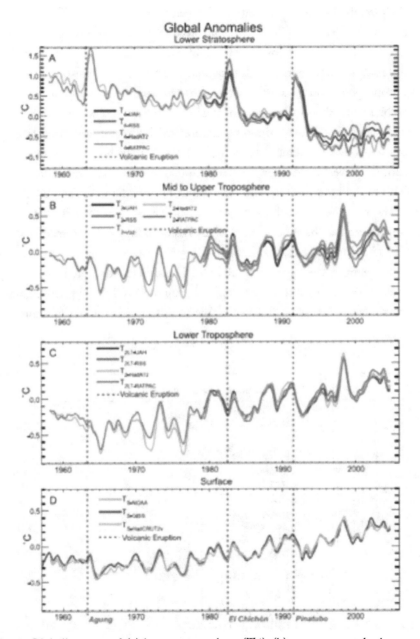

FIGURE 4.3: Globally averaged (a) lower stratosphere (T4), (b) upper tropospheric temperature (T2), middle tropospheric temperature (T*G), and (d) lower tropospheric temperature (T2LT). UAH (blue), RSS (red), and UMd (black) from satellites. HadAT2 (green) radiosonde data.

Table 4.1 shows the location of the layers. The teams are identified in Table 4.2. It is important to note that both Spencer and Christy were involved in producing the report.[8]

Figure 4.3a shows that the stratosphere has cooled substantially. This is in accord with radiation theory and climate model simulations. If increased solar forcing were responsible for the warming of the lower troposphere and the Earth's surface, the temperature of the stratosphere would have warmed. The second panel shows the region of the lower stratosphere and the upper troposphere. Because the stratosphere has cooled, the temperatures in this region are more or less flat. The third panel, however, shows the temperatures in the lower troposphere. A considerable warming is evident. Finally, the lowest panel shows the surface temperature data from three data sets. The comparison of this to the lower troposphere data is striking. The satellite data reveal temperatures in thick layers of the atmosphere. Radiosondes are temperature sensors carried aloft by weather balloons. Local temperature data have been collected by radiosondes routinely since 1958. Unlike the satellite data, the radiosonde data are sparse; that is, they do not cover large areas and they measure only the vertical temperature changes over small areas. The available data are in agreement with the satellite data. They show that, except for the tropics during the 1979–2004 period, the troposphere temperature is increasing, whereas in all data sets, the stratosphere is cooling substantially.

TABLE 4.1: Where the different temperatures are located

Troposphere (satellite)	T^*_G	Troposphere	Sfc – 13 km	Sfc – 150 hPa
Tropical Troposphere (satellite)	T^*_T	Troposphere (tropics only)	Sfc – 16 km	Sfc – 100 hPa
Mid Troposphere to Lower Stratosphere	T_2	Mid and Upper Troposphere to Lower Stratosphere[2]	Sfc – 18 km	Sfc – 75 hPa
Lower Stratosphere (satellite)	T_4	Lower Stratosphere	14 – 29 km	150 – 15 hPa
Lower Stratosphere (radiosonde)	$T_{(100-50)}$	Lower Stratosphere	17 – 21 km	100 – 50 hPa

2 Only about 10% of this layer extends into the lower stratosphere.

TABLE 4.2: Teams involved in reconstruction of atmospheric temperature trends

Our Name Web Page	Name Given by Producers	Producers
Surface		
NOAA http://www.ncdc.noaa.gov/oa/climate/monitoring/graph/graph.html	ER-GHCN-ICOADS	NOAA's National Climatic Data Center (NCDC)
NASA http://www.giss.nasa.gov/data/update/gistemp/graphs/	Land+Ocean Temperature	NASA's Goddard Institute for Space Studies (GISS)
HadCRUT2v http://www.cru.uea.ac.uk/cru/data/temperature	HadCRUT2v	Climatic Research Unit of the University of East Anglia and the Hadley Centre of the UK Met Office
Radiosonde		
RATPAC http://www.ncdc.noaa.gov/oa/cab/ratpac/	RATPAC	NOAA's Air Resources Laboratory (ARL), Geophysical Fluid Dynamics Laboratory (GFDL), and National Climatic Data Center (NCDC)
HadAT2 http://www.hadobs.org/	HadAT2	Hadley Centre of the UK Met Office
Satellite		
Temperature of the Lower Troposphere		
T_{2LT}-UAH http://vortex.nsstc.uah.edu/data/msu/t2lt	TLT	University of Alabama in Huntsville (UAH)
T_{2LT}-RSS http://www.remss.com/msu/msu_data_description.html	TLT	Remote Sensing System, Inc. (RSS)
Temperature of the Middle Troposphere		
T_2-UAH http://vortex.nsstc.uah.edu/data/msu/t2	TMT	University of Alabama in Huntsville (UAH)
T_2-RSS http://www.remss.com/msu/msu_data_description.html	TMT	Remote Sensing System, Inc. (RSS)
T_2-UMd http://www.atmos.umd.edu/~kostya/t2ch	Channel 2	University of Maryland and NOAA/NESDIS (UMd)
Temperature of the Middle Troposphere minus Stratospheric Influences		
T^*_{G} (global) T^*_{T} (tropics) http://www.ncdc.noaa.gov/oa/climate/research/lo-mt-uah-monthly-anom.txt (UAH) http://www.ncdc.noaa.gov/oa/climate/research/lo-mt-rss-monthly-anom.txt (RSS)	$T_{(850-300)}$	University of Washington, Seattle (UW) and NOAA's Air Resources Laboratory (ARL)
Temperature of the Lower Stratosphere		
T_4-UAH http://vortex.nsstc.uah.edu/data/msu/t4	TLS	University of Alabama in Huntsville (UAH)
T_4-RSS http://www.remss.com/msu/msu_data_description.html	TLS	Remote Sensing System, Inc. (RSS)
Reanalysis		
NCEP50 http://wesley.wwb.noaa.gov/reanalysis.html	NCEP50	National Centers for Environmental Prediction (NCEP), NOAA, and the National Center for Atmospheric Research (NCAR)
ERA40 http://www.ecmwf.int/research/era	ERA40	European Centre for Medium-Range Weather Forecasts (ECMWF)

Taken together, the data show the following:

1. The globally averaged surface temperature increased at a rate of about 0.12 °C per decade since 1958 and about 0.16°C per decade since 1979.
2. Globally averaged tropospheric temperature increased at a rate of about 0.14°C per decade since 1958 according to the radiosonde data.
3. The reanalyzed MSU data are consistent with the surface data, exhibiting a warming of about 0.12°C to 0.16°C per decade since 1979.

Quoting directly from the executive summary of the report:

Since the late 1950s, all radiosonde data sets show that the low and mid troposphere have warmed at a rate slightly faster than the rate of warming at the surface. These changes are in accord with our understanding of the effects of radiative forcing agents on the climate system and with the results from model simulations.

For observations during the satellite era (1979 onwards), the most recent versions of all available data show that both the low and mid troposphere have warmed. The majority of these data sets show warming at the surface that is greater than in the troposphere. Some of these data sets, however, show the opposite—tropospheric warming that is greater than that at the surface. Thus, due to considerable disagreements between tropospheric data sets, it is not clear whether the troposphere has warmed more or less than the surface.

The most recent climate model simulations give a rand of results for changes in globally averaged temperature. Some models show more warming the troposphere than at the surface, while a slightly smaller number of simulations show the opposite behavior. There is no fundamental inconsistency among these model results and observations at the global scale.

Studies to detect climate change and attribute its causes using patterns of observed temperature change in space ant time show clear evidence of human influences on the climate system (due to changes in greenhouse gases, aerosols, and stratospheric ozone).

The observed patterns of change over the past 50 years cannot be explained by natural processes alone, nor by the effects of short-lived atmospheric constituents (such as aerosols and tropospheric ozone) alone.

The bottom line here is enlightening. Because the satellite data show about the same temperature trends as the surface data and especially because the variations, the wiggles in the temperature curves are so strikingly similar, it gives us much more confidence that the surface temperature measurement is accurate!

Challenge 8. *The urban heat island effect has contaminated the surface temperature record and exaggerated the appearance of warming.*

There have been a number of claims, most recently by Peterson, that urban heat islands are largely responsible for the measured increase in global temperatures.[10] Many studies have shown that urban areas are generally warmer than nonurban areas because they absorb more solar heat. First of all, the satellite data agree well with surface data, as shown in Figure 4.3, and it covers the entire Earth, 70% of which is covered by oceans and perhaps 90% is not compromised by heat islands. Second, the temperature rise over ocean areas is similar to that over land areas. Third, studies (e.g., Peterson[10]) that have looked carefully at temperature on hemispheric and global scales (i.e., averaged over large areas) conclude that any urban-related trend is very much smaller than decadal or longer term trends. This could be partly because some stations with obvious urban-related warming trends have been omitted from the global data sets and also because, in any case, many urban measuring stations are located in parks, where urban warming effects are smaller.

Finally, Parker, in two recent articles in *Nature*[11] and the *Journal of Climate*[12], noted that warming trends in night minimum temperatures during the 1950–2000 period were not enhanced on calm nights, which would be the time most likely to be affected by global warming. Thus, the global land warming trend is very unlikely to be influenced significantly by increasing urbanization. The most-recent report by the IPCC estimates the uncertainty from urban warming to be as small as 0.002°C per decade.

Thus, the claim that urban warming has compromised the data on global warming is not valid.

Challenge 9. *What about that Greenpeace questionnaire?*

One of the strangest claims by the global warming skeptics is that most climate scientists are also skeptical of global warming. The general claim is that only 13% of climate scientists subscribe to the alarming version of greenhouse warming, whereas 32% said that the theory was possible and 47% said that it is probably not true. To make things even more interesting, these responses were gleaned from a questionnaire by the "radical" environmental group Greenpeace. It is perhaps convenient that the skeptics do not usually state the question posed by Greenpeace. Yet, to understand the implications of the answers, one must know precisely what the questions were. The first question was "Do you think there will be a point-of-no-return, at some time in the future, at which continued business-as-usual policies run a serious risk of instigating a runaway greenhouse effect?" What is a runaway greenhouse effect? For the benefit of readers who might be unfamiliar with the scientific jargon, a "runaway greenhouse effect" means that, even if we were to stop producing greenhouse gases, internal feedbacks within the climate system itself would continue to produce greenhouse gases until the planet became as hot as, say, Venus, and therefore uninhabitable. This is indeed an extreme view. I am surprised that nearly half of the scientists questioned responded that this is possible. I am among the 47% who believes this is not likely. However, the belief that a runaway greenhouse effect is improbable does not imply a belief that greenhouse-induced climate change will not be severe. The next question posed by Greenpeace was "How satisfied are you with the progress at the climate negotiations?" Sixty-two percent responded that they were too slow or far too slow. Only 8% responded that the progress was too fast or far too fast. The third question posed by Greenpeace was "How seriously has the work of climate scientists been taken since the IPCC report (of 1992) and the Scientists' Statement of the World Climate Conference were published in 1990?" Only 3% responded that this was given too much or far too much weight. The rather clear implication is that nearly half the scientists questioned (judging by the response to the first question) believe that the Scientists' Statement of the World Climate Conference did not go far enough. Practically all the scientists who responded believe that the Scientists' Statement, that is, we face substantial climate change due to a human-induced greenhouse effect, ought to be taken seriously.

Challenge 10. *What about those ice age predictions?*[13]
A great deal has been made about a few books that appeared a few years ago predicting that we were coming to the end of the current interglacial and beginning to enter the next ice age. All of these books were written decades ago. I have seen three such books, and two of them also discussed the possibility of greenhouse warming. In other words, the authors were saying that something was in the works and that the climate might go in either direction. Citing this as a reason to doubt current model predictions and data is sort of a "nya-nya-nya-nya-nya" argument. In any case, the ice age predictions were based largely on the conjecture that the present interglacial has lasted about as long as past ones and that we are sort of "due." The greenhouse effect is based on hard scientific facts about the increase of CO_2 in the atmosphere and how the climate reacts to the increase in greenhouse gases.

Challenge 11. *Major Arctic outbreaks have occurred in places such as Florida in recent years (Michaels, Sound and Fury, pp. 68–70).*[2]
Although there is little doubt that the globally averaged temperature of the Earth has increased during the past century, it is true that, even during warm years, there have been unusual cold spells

in the southern United States. How can this be? Nobody has said that every day would be warmer than usual on a warmer Earth. In fact, one of the fears of climatologists is that variability in the week to week and month to month as well as decade to decade weather and climate might increase. Indeed increased variability can wreak more havoc than the warming itself. Late frosts and extreme rainfall events as well as unusually cold winters or warm summers can deal death blows to farmers. Why should variability increase as the planet warms? Nobody can say for sure, but there are some conjectures. There is almost irrefutable evidence that during warmer climates high latitudes warm more than middle and low (equatorial) latitudes. We know this because, during warmer periods in the past high, latitudes warmed more than low latitudes, and during colder periods (like the last glacial age), high latitudes cooled more than low latitudes. It follows that whatever physical mechanism is responsible for the transport of heat from the warm tropics to the cold polar regions must be more active during warm periods. One mechanism is the formation of fronts that generate the local weather patterns as described earlier. Thus, cold fronts might be expected to be more severe and reach further equatorward during globally warmer periods. Although this is physically reasonable, it is, of course, an unproven conjecture. But it is worth stating that this is exactly what happens during El Niño years. The fact that there are cold spells in Florida, in any case, in no way implies that the Earth is not warming. It is.

Challenge 12. *The ocean temperatures at intermediate depths increased between 1950 and 1980, whereas the ocean surface temperature decreased. No cause is "suggested or implied" for this. "An explanation would probably revolutionize our understanding of the problem."*[2]

This statement is taken from Patrick Michaels's book, *Sound and Fury* (pp. 85–87).[2] This surprising statement indicates that Dr. Michaels may not be familiar with the climate literature! The climate literature is so vast that it would not be surprising if every climate scientist is not familiar with all of the published work. But one of the premier journals in the climate literature, *Climatic Change*, has published a plausible explanation to the phenomenon to which Michaels refers,[14] not one but three times! It has also been published in the prestigious *Journal of Geophysical Research* several times.[15] It has to do with the formation of ocean deep water, as illustrated in Figure 1.4. It has been shown in several articles that if the rate of sinking of the salty and cold water in the North Atlantic were to decrease even slightly, heat diffusing from the surface layer of the ocean would cool the surface and warm the deeper water. Warming a layer of ocean water 1000 m thick by just two tenths of a degree centigrade requires that an enormous amount of heat leak out of the upper layer of the ocean. If this happened in just more than 15% of the world's oceans during the 30-year span between 1950 and 1980, all of the extra heat due to greenhouse warming during this period would simply go to warming the deep water. None would be left to warm the surface. This may very well be the reason why the northern hemisphere surface temperature stopped warming during this period! This is not an airtight explanation, but it is certainly plausible, and it supports the idea that the Earth is warming. It also implies that if and when the rate of sinking of cold water increases again, the surface water will warm fairly rapidly.

Challenge 13. *The data do not matter.*[2]
This is a statement supposedly made by Folland at a conference on global warming and was made fun of by Patrick Michaels in his book, *Sound and Fury*. Taken out of context, it sounds as if Folland thinks that whenever data do not agree with theory, one should believe the theory and simply reject the data. In fact, Folland made the remark in response to an argument that I have heard Michaels make on more than one occasion. The argument goes like this. Before 1950, the concentrations of greenhouse gases in the atmosphere were rising slowly, but the global temperature was rising fairly rapidly. After 1950, when emissions of CO_2 were rising rapidly, the global temperature actually cools for a while. This is especially true in the northern hemisphere. The southern hemisphere continues to warm slightly, but the rate is smaller than it was before 1950. Michaels then states that "the northern hemisphere should be the first and fastest to warm" because the fraction of the surface covered by oceans is larger in the southern hemisphere. What Folland was apparently attempting to explain to Michaels was that both the natural variability (perhaps, variations in the North Atlantic Ocean as explained in the above paragraph) and the presence of sulfur aerosols (almost all in the northern hemisphere) could easily explain the dip in the northern hemisphere temperature. The data displayed in this figure certainly do not cause one to retreat from the fundamental science.

Challenge 14. *Reducing CO_2 emissions by 60% to 80% is necessary to stabilize CO_2 in the atmosphere.*[16]
Several of the skeptics present this bit of information, but for what reason is not clear. Perhaps, we should this information, which is true, admit that such a reduction is not feasible, and conclude either that either global warming is not real or that we shall surely have to learn to live with it. An alternative is to view it as a very great challenge and, if it is indeed occuring, we had best begin thinking seriously about how to meet it. In any case, essentially everyone agrees that this statement is true, whatever their position on global warming.

Challenge 15. *Will sea level rise? If so, by how much, and how rapidly?*[2]
Michaels says that it is plausible that the polar ice caps could become thicker, making sea levels fall.[2] As we have seen, sea level has been rising since the end of the last glacial cycle, and there has been an apparent increase in the rate of rise during the last 100 years. Sea level rise results from two phenomena: the thermal expansion of sea water as it is heated and the melting and runoff into the ocean of land ice. Since the thermal expansion of cold sea water is small, any sea level rise must be caused by an increase in the temperature of the warm part of the ocean. This is virtually a certainty

as the climate warms. We have seen that glaciers virtually everywhere are melting fairly rapidly, contributing their part to sea level rise (Figure 3.14). The two largest pieces of land ice are the Greenland and Antarctic ice sheets. As the climate warms, atmospheric moisture tends to penetrate to higher latitudes, where, unless the high latitudes warm more than is usually predicted, it might well fall as snow, building up the ice sheets, and contributing to a fall in sea level. GCMs probably do not correctly predict the poleward exaggeration in the amplitude of the warming, and in the real-world future, the high latitude warming might just be enough to begin melting the Greenland ice sheet and eventually the Antarctic ice sheet. In fact, as we saw in Chapter 3, the flow speed of several Greenland and Antarctic outlet glaciers, which drain ice from the interior of the ice sheets, has recently increased substantially. Satellite measurements show that the amount of ice melt flowing into the sea from large glaciers in southern Greenland has more than doubled in the last 10 years. A new study in the journal *Science* suggests that sea levels will rise by 20 to 55 inches by 2100. If Greenland should melt entirely, sea level would rise by about 7 m (23 ft). If West Antarctica melts, sea level would rise by 5 m (17 ft), and if the rest of Antarctica should melt, sea level would rise about 77 m (253 ft). There is particular concern about the West Antarctic ice sheet, which some scientists believe is almost unstable already. This is, I believe (at least for the reasonably near future, say, the next 50 years), a low-probability catastrophic event, but if it were to happen, I would not like to be living near an ocean! I do live near the Gulf of Mexico, and living in New Orleans may become problematic.

The claims of the skeptics are thus soundly debunked. The only claim that is so far controversial is the cosmic ray theory. Together with the lack of any proven physical mechanism, the recent article by Lockwood and Frohlich, and the plausibility of the other causal factors, it seems highly unlikely that global warming is caused by any mechanism other than the increase in atmospheric greenhouse gases, the most prominent being the increase in atmospheric CO_2.

In fact, the skeptics whom I know personally are now admitting that global warming is happening, and happening in response to increasing greenhouse gases, the most important being CO_2. Their response has now become: it would not warm as much as is predicted. And so what; it would not be harmful and it might even be a good thing. Let us explore those ideas.

NOTES AND REFERENCES

Figures 4.1 and 4.2 are from the IPCC Fourth Report of the Working Group I.
Figure 4.3 is from Reference 7.

1. Raymond, L.R., chairman and chief executive officer, Exxon Corporation, in a speech given at the World Petroleum Conference, Beijing, China, October 19, 1997.
2. Michaels, P., *Sound and Fury: The Science and Politics of Global Warming*, The Cato Institute, Washington, DC, 1992.

3. Jastrow, R., Nierenberg, W., and Seitz, F., "*Global Warming: What does the Science Tell us?,*" George. C. Marshall Institute, Washington, DC, 37 pp., 1990.

4. Wang, Y.M., Lean, J.L., and Sheeley, N.R., "Modeling the sun's magnetic field and irradiance since 1713," *Astrophysics Journal*, vol. 625, pp. 522–538, 2005.

5. Svensmark, H., "Cosmoclimatology: a new theory emerges," *Astronomy & Geophysics*, vol. 48, issue 1, pp. 1.18–1.24, 2007.

6. Lockwood, M., and Frohlich, C., "Recent oppositely directed trends in solar climate forcing the global mean surface air temperature," *Proceedings of the Royal Society A*, vol. 1880, pp. 1–14, 2007.

7. Carslaw, K. S., Harrison, R.G., and Kirby, J., "Cosmic rays, clouds and climate," *Science*, vol. 298, pp. 1732–1737, 2002.

8. Spencer, R.W., and Christy, J.R., "Precise monitoring of global temperature trends from satellites," *Science*, vol. 247, pp. 1558–1562, 1990.

9. Karl, T.R., Hassol, S.J., Miller, C.D., and Murray, W.I., Eds., *Temperature trends in the lower atmosphere: steps for understanding and reconciling differences. A report by the Climate Change Science Program and Subcommittee on Global Change Research*, Washington, DC, 2006.

10. Peterson, T.C., "Assessment of urban versus rural in situ surface temperatures in the contiguous United States: no difference found," *Journal of Climate*, vol. 16, pp. 2941–2959, 2003

11. Parker, D.E., "Large-scale warming is not urban," *Nature*, vol. 432, pp. 290–290, 2004.

12. Parker, D.E., "A demonstration that large-scale warming is not urban," *Journal of Climate*, vol. 19, pp. 2882–2895, 2006.

13. For example, see Ponte, L., *The Cooling*, Prentice-Hall, Indianapolis, IN, 1976; anonymous, *The* Weather Conspiracy: The Coming of the New Ice Age, Bantam Books, New York, 1977; Gribbin, J., *Forecasts, Famines and Freezes*, Pocket Books, New York, 1977.

14. Watts, R.G., and Morantine, M.C., "Is the greenhouse gas–climate signal hiding in the deep ocean?," *Climatic Change*, vol. 18, pp. iii–vi, 1991.

15. Watts, R.G., "Global climate variations due to fluctuations in the rate of deep water formation," *Journal of Geophysical Research*, vol. 90, pp. 8067–8070, 1985.

16. For example, see Hoffert, M.I., Caldeira, K., Jain, A.K., Haites, E.F., Harvey, L.D.D., Potter, S.D., Schlesinger, M.E., Schneider, S.H., Watts, R.G., Wigley, T.M.L., and Wuebbles, D.J., "Energy implications of future stabilization of atmospheric CO_2 content," *Nature*, vol. 395, pp. 881–884, 1998.

· · · ·

CHAPTER 5

Impacts: The "So What" Question

This is where the water gets a little muddy. There are some who believe that any climate change, whether it gets warmer or cooler, will have a negative overall impact. I am among these. The reason is that civilization, along with plants and animals, has adapted to the current climate. There is a huge infrastructure in place supporting agriculture where it can currently be successfully practiced, for example. Climate change will lead to weather patterns different from those of today. Some regions will become dryer, and some will become wetter. Although regions where crops do not presently grow might well become better suited for farming, this is not comforting to farmers whose land will become less suitable. (They could sell their farms and move to better regions, but to whom would they sell?). The reader should also be reminded that summer soil moisture is predicted to decrease virtually everywhere by all of the climate models. If the warming occurs during a relatively short time span (decades to a century), which it probably will, relocation can become a huge problem. If sea levels rise at rates even close to those predicted by some models,[1] relocation of people from countries such Bangladesh, from island nations, and from coastal cities could be a problem of almost unimaginable proportions.

We have seen that climate change has occurred in the past simply because of "natural variability," that is, variations in the climate brought about mostly by its internal workings and not principally through the influence of humans. It is interesting to note that many of these rather small variations have had rather large impacts on people and civilizations, especially locally and regionally.[2] For example, during the Medieval Warm Period, grapes flourished in England, and Greenland was settled by Europeans. These events were interrupted by the Little Ice Age. There is evidence that the cool period that seems to have begun about 4500 years ago adversely affected a number of civilizations. Before about 2300 B.C., a region of northern Mesopotamia between the Tigris and Euphrates rivers called the Habur Plains was rich in agriculture. Sustained by this highly productive agricultural area, the cities of this region, where the written word, metalworking, and bureaucracy were probably born, thrived. At about 2200 B.C., a slight global cooling appears to have brought with it a drought in the region that crippled this early civilization. Moreover, the drought coincided with the collapse of perhaps half a dozen city-based civilizations in that part of the world—in Egypt, Palestine, Crete, Greece, and in the Indus Valley. In none of these cases was the climate change as large as that which is expected to occur during the next century from global warming.

5.1   STORYLINES

At the International Institute for Applied Systems Analysis, Nakicenovic and Swart have developed scenarios, or storylines, to have something against which to test how climate will likely change in the future. The only thing that is certain to be true about scenarios for the future is that they are all going to prove to be wrong to a greater or lesser degree. Nevertheless, it does help to examine "what-if" stories to get some idea of what the future might hold. The four storylines developed at International Institute for Applied Systems Analysis are presented as two sets. Set A looks at the storyline emphasizing strong economic values and another emphasizing strong environmental values. Set B examines increasing globalization versus increasing regionalization. In simple terms, the four storylines, as described in the IPCC Fourth Report, are (quoting that document) as follows:

A1: A future world of very rapid economic growth, global population that peaks in midcentury and declines thereafter, and rapid introduction of new and more efficient technology.

A2: A very heterogeneous world with continuously increasing global population and regionally oriented economic growth that is more fragmented and slower than in other storylines.

B1: A convergent world with the same global population as in the A1 storyline but with rapid changes in economic structures toward a service and information economy, with reductions in material intensity, and the introduction of clean and resource-efficient technologies.

B2: A world in which the emphasis is on local solutions to economic, social, and environmental sustainability, with continuously increasing population (lower than A2) and intermediate economic development.

The A1 storyline is also developed into three groups that describe alternative directions of technological change in the energy system. The three A1 groups are distinguished by their technological emphasis: fossil fuel intensive (A1F1), nonfossil energy sources (A1T), or a balance across all sources (A1B), where "balanced" is defined as not relying too heavily on one particular energy source, on the assumption that similar improvement rates apply to all energy supply and end use technologies. To see the scenarios in much more detail, see Appendix 5.1.

In my opinion, all of these are a little optimistic. Even the so-called business-as-usual scenario outlined in an early report of the IPCC, which is labeled IS92a, assumes that, by 2050, some 25% of global energy will come from renewable sources. This is about 7,500,000,000,000 W of energy, or about half as much as the entire world now consumes!

5.2   MODEL PREDICTIONS

Let us now look at some of the climate model predictions for the 21st century. Multimodel predictions are used. These are assemblies of several models to construct a kind of prediction that most models agree on.

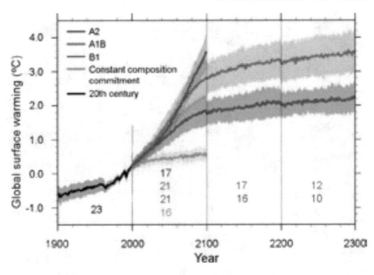

FIGURE 5.1: Multimodel means of surface warming relative to 1980–1999 for scenarios A2, A1B, and B1, shown as continuation of the 20th century simulation. Lines show the multimodel means, and the shading denotes ±1 standard deviations of individual model runs. The colored numbers represent the number of models used for the simulations. (One standard deviation means that 68% of the results are within the shaded area.)

Figure 5.1 shows the predicted globally averaged surface temperature rise for several of the scenarios described above. It is worth noting that the differences between the results are small for the next couple of decades. The orange curve reflects the temperature increase if CO_2 is stabilized immediately. The temperature continues to rise because the CO_2 that is already in the atmosphere will remain for centuries. By midcentury, the other three model groups show temperature rises from 1900 of between 1.75°C and 3°C.

Figure 5.2 shows predicted regional temperature changes for three time periods within the 21st century. Note that the high latitudes in both hemispheres warm significantly more than mid and low latitudes. Arctic temperatures are 7.5°C to 8°C warmer than the period 1980–1999.

Predictions of changes in precipitation and soil moisture for the 2080–2099 period relative to the 1980–1999 period are shown in Figure 5.3 following scenario A1B, which is the midrange scenario for warming, as can be seen from Figures 5.1 and 5.2. One thing to notice in the figure is that there are some places, for example, parts of China and northern India, where, although precipitation increases, soil moisture decreases because of increased evaporation. This also shows up in parts of western South America, particularly in the Amazon basin.

With these figures in mind, let us discuss some of the likely impacts of global warming.

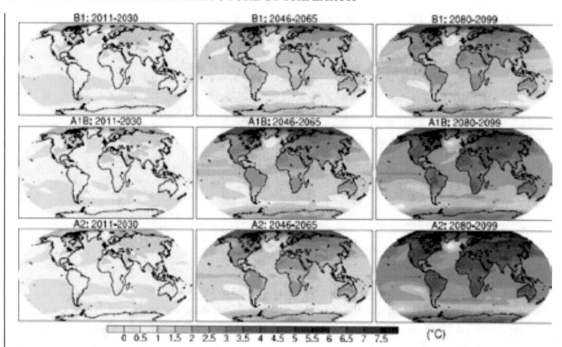

FIGURE 5.2: Multimodel mean of mean annual surface warming in centigrade for scenarios B1, A1B, and A2 for the three average periods 2011–2030, 2046–2065, and 2080–2099. Warming is relative to the average of the 1980–1999 period.

5.3 DROUGHTS

Climate models have, in essentially every case, correctly predicted the droughts that are occurring today. There is every reason to believe that future droughts will follow these patterns and that the multimodel predictions have considerable credibility.

Fresh water will be in ever-shorter supply in many places as the climate changes. While responding to increasing temperatures, river flows may increase overall, with some rivers becoming more swollen, but many of those that provide water for most of the world's people will begin to dry up.

Increases in precipitation will cause the discharge of fresh water from some rivers around the world to rise by almost 15%. However, at the same time, water stresses are predicted to increase significantly in regions that are already relatively dry. Evaporation will reduce the moisture content of soils in many semiarid parts of the world, including northeast China, the grasslands of Africa, the Mediterranean, and the southern and western coasts of Australia. Soil moisture will fall by up to 40% in southern states of the United States. The Amazon basin is also predicted to suffer increasing drought.

Some of these predicted changes are already happening. One study shows that temperature changes have affected the flow in many of the world's 200 largest rivers during the past century, with the flow of Africa's rivers declining during the past 10 years. Climate change during the past few decades has already caused discharge from some rivers in North and South America and Asia to increase. But some have decreased substantially (see below). Runoff in Europe has remained stable, but the flow of water from Africa's rivers has fallen.

5.3.1   AFRICA

Failing rains are already a major cause of hardship in Africa, as witnessed by the current drought in East Africa, and the results shown in Figure 5.3 certainly imply that global warming will change rainfall patterns across the continent.

Rainfall has a direct effect on the way water drains into streams and rivers and, therefore, on supply. When rainfall in an area is less than 400 mm/yr, there is virtually no drainage into rivers and streams. Therefore, a decrease in precipitation will amplify the loss of available water because evaporation is increased. A 10% decrease in rainfall could cut the available water by 50%. Semiarid regions in Africa, such as southern Africa and the Sahel, are extremely vulnerable to the spread of deserts.

5.3.2   INDIA

With a long dry spell ruining crops in northwest India, conditions in the country's breadbasket are the worst in a decade. Among the affected areas are the major rice- and cereal-growing states of Haryana, Punjab, and Rajasthan. In India's largest state, Uttar Pradesh, state officials declared 21 districts as drought-hit and sent drinking water to the driest areas.

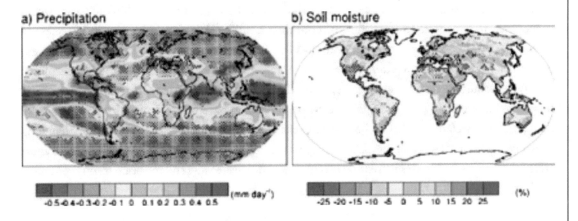

FIGURE 5.3: Multimodel-predicted changes in precipitation (mm/day) and soil moisture content (%). Regions are stippled where the various models agree at least 80% on the sign of the change. Changes are for scenario A1B for the 2080–2099 period relative to the 1980–1999 period.

The monsoon season is crucial to India's farm output. About 80% of the country's rainfall occurs between June and September. Two thirds of India's 1 billion people depend directly on agriculture for their livelihood, and farming contributes about 25% to gross domestic product (GDP).

Although droughtlike conditions persisted in some areas, thousands of mud and thatch houses were washed away in the eastern state of Bihar and in northern Assam after heavy downpours. According to Figure 5.3, precipitation is set to increase, especially in northern India, whereas soil moisture continues to decrease.

5.3.3   SOUTH AMERICA

Widespread drought in 2005 in the Amazon basin was the worst in some areas since records began and prompted the Brazilian government to declare a state of emergency. There is broad consensus that the 2005 drought was linked not to El Niño, the periodic phenomenon that begins with a warming of Pacific waters off the coast of South America, as was the case with most previous droughts in the Amazon, but to warming sea surface temperatures in the tropical North Atlantic. At least one climate model predicts that, under current levels of greenhouse gas emissions, the chances of such a drought would rise from 5% now (one every 20 years) to 50% by 2030 and to 90% by 2100. The multimodel prediction in Figure 5.3 indicates substantial decreasing soil moisture despite increasing precipitation.

Downstream of the Amazon basin drought area, in the city of Manaus, the level of the Amazon River dropped by 3 m during the current drought. Many communities dependent on the river for transport were left stranded as tributaries dried up.

5.3.4   CHINA

In China, the devastating drought that has plagued the drainage area of the Yellow River, which is supposed to supply irrigation water to more than a million hectares (a hectare is 2.7 acres) of farmland, shows no signs of abating. More than 300 000 farmers are now in urgent need of irrigation water. Farmers in the region are being encouraged to plant crops that need less water. Even if these measures are taken, a shortage of more than half a billion cubic meters will remain. More than 48 million people and 5 million hectares of arable land have been affected in some way. More than 3 million hectares yielded no harvest last year. The central and western parts of the Shandong Province, including Heze, Jining, Liaocheng, Jinan, Zibo, Binzhou, and Dongying, are the most affected. Figure 5.3 indicates increasing precipitation and decreasing soil moisture throughout most of China.

5.3.5   AUSTRALIA

Water reservoirs in Australia are dropping fast, crop forecasts have been slashed, and great swaths of the continent are entering what scientists have called a "one-in-a-thousand-years drought." Many regions are in their fifth year of drought, and more than half of Australia's farmland is experiencing drought. The Murray–Darling river system, which receives 4% of Australia's water but provides three quarters of the water consumed nationally, is already 54% below the previous record minimum. The drought will affect drinking water supplies to many areas. Sydney's largest reservoir is now only 40% full, and many small rural towns in east Australia face shortages. Thus, a serious impact on crops is expected: its lowest wheat crop for 12 years, a decrease of more than 50% below last year's, which was already well below average.

The Commonwealth Scientific and Industrial Research Organization, the leading scientific body in Australia, has predicted that rainfall in parts of eastern Australia could drop by 40% by 2070, along with a 7°C rise in temperature. Figure 5.3 shows decreasing precipitation except in the north and decreasing soil moisture, especially on both coasts.

5.4   AGRICULTURE AND WORLD FOOD SUPPLY

Obviously, predictions concerning the future of agricultural productivity depend on which of the various scenarios one uses. The ability of civilizations and agriculture systems to adapt to climate change also depends on such things as population growth, access to information, farming technology, and the ability of the people and the government to afford irrigation and fertilizer, among many other things. It also depends on the crop varieties that are grown in different locations and on the crop varieties that are adapted to these locations. The least-developed countries are most vulnerable to climate change because of their limited capacity to cope with change, particularly extreme events such as floods and drought. Also, most developing countries (apart from the oil-producing countries in the Middle East) have agriculture-based economies. Results of model predictions of all scenarios show a poleward shift in temperature with a significant reduction of arctic ecosystems and increases in arid areas in developing countries.

Some developed countries might see a potential for expansion of land suitable for agriculture, but only if this new land is used for certain crops such as cereals. The potential new land available in North America (much of it in Canada) would be a 40% increase. Because much of it is in the northern latitudes where temperatures will increase and frozen ground will decrease, land suitable for agriculture in the Russian Federation could increase by as much as 60% and in East Asia by 10%. As we have seen, however, much of China may be subject to both droughts and floods, and agricultural productivity there will probably decrease, perhaps substantially.

In most developing and poor countries, global warming is predicted to lead to an increase in land with severe constraints on agricultural productivity. In northern Africa, an increase of as much as 5% could occur by 2080 due to expansions of both the Sahara and Sahel deserts.

According to all the scenarios, developing countries will suffer a large reduction of wheat-producing potential by 2080. Wheat-producing potential is predicted to decrease by 12–27% in South America and by 10–95% in southeast Asia. Wheat virtually disappears from Africa.

Because CO_2 is what plants "eat," one might think that an increase in atmospheric CO_2 would be a boon to agriculture. This is referred to as CO_2 fertilization. When some plants are exposed to increased CO_2, they grow better, and in addition, their stomatal openings (small openings in the leaves that admit CO_2 and water vapor) close, reducing their loss of water vapor. They, therefore, can survive under reduced water availability. C3 plants, which are most eaten crops except corn, sugar cane, sorghum, and millet, appear to thrive on increased CO_2. On the other hand, C4 plants do not seem to respond nearly as positively.[4] C3 and C4 plants "digest" carbon from CO_2 through different biological pathways, and that is all you need to know at this stage.

Weeds will probably benefit from CO_2 fertilization. Eighty of the 86 plant types that contribute 90% of global food supplies are C3 plants, whereas 14 of the 18 most destructive weeds are C3 plants.[5]

Some studies question whether CO_2 fertilization has a very large effect, even on C3 plants. There is also evidence that the protein concentration is reduced in plants growing in high CO_2, thereby reducing their nutritional value.[6] Also, experimental studies are generally performed in environmental chambers under ideal conditions of the availability of other nutrients and water. When models are used, they are designed to mimic these conditions. However, the 19th-century German chemist Justis Liebig discovered that the rate at which a plant can grow depends on the supply of the least available nutrient.[7] Plants require nutrients in addition to CO_2, and if one or more of these are in short supply, the effect of CO_2 fertilization may be nil. This may be particularly important for underdeveloped countries, where fertilizers, herbicides, pesticides, and irrigation are either not available or prohibitively expensive. In addition, plants growing in experimental settings have less competition from weeds and pests than those growing in farmers fields.

It appears that insect pests will prosper during a warmer climate, but this is unclear. The principal concern appears to be with species that can increase their populations by producing an extra generation each year in warmer climates, or increase their range. However, at least some insect species become less fertile under higher temperatures. The decreased nutritional value of the plants might increase insect mortality, but the surviving ones would have to eat more to survive. Warmer temperatures, higher humidity, and more frequent summer precipitation will increase certain plant diseases if these things occur. Mild winters have been connected to outbreaks of powdery mildew, brown leaf rust in barley, and strip rust in cereal crops.[8] Mild winters combined with warm summers provide optimal growth conditions for cercosporia leaf spot disease and potato blight.

Some of the negative effects of global warming can be offset by adaptive measures such as irrigation and fertilizer use. Hence, the industrialized countries are likely to suffer far less than the developing countries, where these measures may be unavailable or prohibitively expensive.

5.5   SEVERE WEATHER EVENTS

Both daily maximum and minimum temperatures are increasing over most land masses, but minimum temperatures are increasing more rapidly than maximum temperatures. The result (in many places) is that the length of the frost-free period has increased by as much as 2 weeks from the 1950s to the 1990s in some regions. Extreme heat waves have increased in frequency, with extreme nighttime temperatures together with certain societal factors (e.g., the absence of air conditioning in the homes of poor people) resulting in the large number of heat-related deaths in the summer of 1995 in Chicago.[9] Nevertheless, there has been an apparent increase in the frequency of major freezes in Florida, even as the mean winter temperature increased. There is some evidence of an increase in climate variability when mean temperatures increase, and this may explain the situation in Florida. This may be the result of a change in storm tracks that occurred in the 1970s and 1980s. Climatologists have examined long-term observations of the frequency of extreme rainfall events by analyzing trends in the percentage of total seasonal and annual rainfall occurring in heavy daily events over various regions of the world.[10] In the United States, Europe, Japan, and parts of Australia, an increasing trend is evident. Data in the United States have revealed a very significant increase in extreme rainfall events in most areas of this country. The increases are evident in both the frequency and intensity of these events. Significant flooding occurs mostly after persistent 2- or 3-day rainfall events because after 2 days, the ground becomes saturated, and runoff into rivers and streams causes them to overflow their banks. (Such an event is presently occurring in Texas and parts of Oklahoma, although the areas are under drought conditions.) Similar data for Canada, the former Soviet Union, and Australia show clear evidence of an increase in extreme rainfall events. There is strong evidence of a recent increase in the frequency or intensity of droughts. It is expected that, as the globe warms, some regions will experience more intense and more frequent droughts. This is because of more rapid evaporation of moisture from plants, soils, and reservoirs. Because of the increased atmospheric moisture, blizzards and snowstorms may actually increase in intensity and frequency.

5.6   TROPICAL STORMS AND HURRICANES

Until quite recently, there was some doubt whether tropical storms were responding to global warming. Now, however, a number of scientists are convinced that the number of super hurricanes has increased as tropical seas become warmer. An online report from the *Philosophical Transactions of the Royal Society of London* on July 27, 2007, stated that the number of tropical storms in the last half of the 20th century was double that in the first half. This certainly seems reasonable because the storms derive their energy from evaporation from the ocean. Figure 5.5 shows a striking correlation between tropical storm frequency and sea surface temperature.

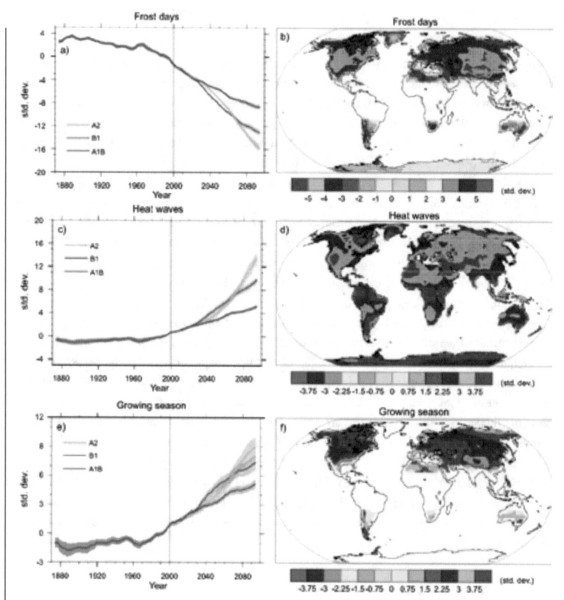

FIGURE 5.4: Changes in extremes based on multimodel simulations. (a) Globally averaged changes in frost day index, defined as number of days per year when the minimum temperature goes below 0°C. (b) Changes in the spatial pattern of frost days between the 1980–1999 and 2080–2099 periods. (c) Globally averaged change in heat waves, defined as the longest period of the year with at least five consecutive days with maximum temperatures at least 5°C higher than the climatology of the same calendar day. (d) Changes in the spatial patterns between 1980–1999 and 2080–2099. (e) Globally averaged changes in growing season length, defined as the length of the period between the first spell of five consecutive days with mean temperature above 5°C and the last such spell of the year. (f) Changes in the spatial pattern between 1980–1999 and 2080–2099.

There is little doubt that as the sea warms, there will be an increase in the intensity of hurricanes.

5.7 THE SEA

The sea may be dying. Phytoplankton, which is the basis of the ocean's food chain, declines when the ocean water gets warmer. Phytoplankton are the microscopic plants the zooplankton and other plant life eat. A NASA satellite tracked water temperature and the production of phytoplankton from 1997 to 2006 and found that when one went up the other went down. The ocean is warming, especially in high northern latitudes and in high southern latitudes. Near Antarctica, waters in summer have warmed in some places by a degree or more, threatening the krill, tiny sea creatures that many organisms depend on for food. In the region near the Antarctica, peninsula krill numbers have fallen by some 80% since the 1970s. This is likely also to be true in the Arctic because krill feed on algae under sea ice, and the arctic sea ice is receding significantly.

Perhaps, more significant is the potential of the changing chemistry of the ocean. More than a third of the CO_2 that is emitted by the burning of fossil fuels is absorbed in the ocean. The

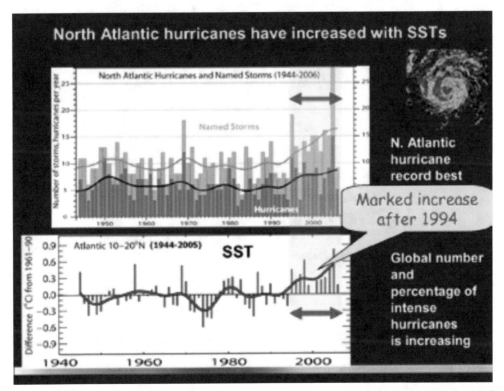

FIGURE 5.5: The number and intensity of Atlantic tropical storms and hurricanes follow closely the tropical Atlantic sea surface temperature.

Intergovernmental Oceanographic Commission reports that 20 to 25 million tons of CO_2 are being added to the ocean each day, and this can only increase as the rate of increase of CO_2 into the atmosphere increases. When CO_2 gas dissolves into the ocean, it produces carbonic acid, decreasing the pH (which means potential for hydrogen) of the ocean water. The pH scale goes from 1 to 14, where lower pH means higher acidity. Too much acidity can lead to disruption of the marine environment. Increasing acidity (decreasing pH) reduces the availability of calcium carbonate, which corals and sea creatures with hard shells rely on to form their shells. It may also affect the growth rates and reproduction rates of fish, and it affects the plankton populations that the fish

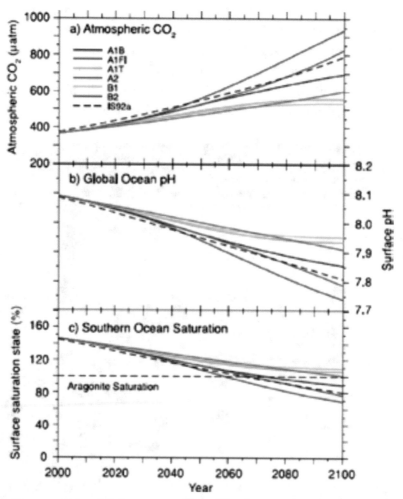

FIGURE 5.6: Changes in the global average surface pH and saturation state with respect to aragonite in the southern ocean under various SRES (Special Report on Emissions Scenarios of the Intergovernmental Panel on Climate Change) scenarios.

rely on for food, with potential disastrous results for the marine food web. The increasing ocean temperatures and the decreasing pH (increasing acidity) of the ocean could impose a sort of double whammy on the ocean.

The pH of the ocean has been about 8.2 for at least many thousands if not millions of years. During the last few decades, it has decreased by about 0.1 unit, and it is projected to decrease faster in the future, as shown in Figure 5.6. Scientists believe that if it reaches 7.6, shell creatures will be unable to form their shells.[11,12] Calcifying organisms such as phytoplankton and zooplankton are the most important players at the base of the ocean food chain.

Boris Worm, an assistant professor of marine conservation biology at Dalhousie University in Halifax, Canada, and his associates have reported an alarming decrease in the number fish species.[13] Some of this is the result of overfishing, but there is little doubt that some reflects the decreasing biodiversity of the ocean and the decrease of the availability of such species as krill that are at the bottom of the food chain, and this is clearly affected by the increase in the temperature of the sea and the disappearance of sea ice that contains the food source of the krill (Figure 5.7).

5.7.1   Coral Reefs

Coral reefs only thrive under very specific conditions because of the symbiotic relationship they have with dinoflagellates, an algae that needs very specific conditions to live in. Less than 0.2% of the global oceans are covered in tropical reefs. Two thirds of the coral reefs are now severely damaged, a fifth so overwhelmingly that they are unlikely to recover. As discussed above, CO_2 absorption could

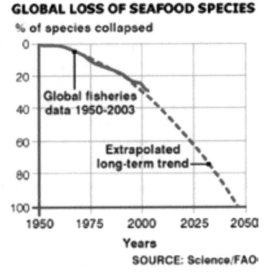

FIGURE 5.7: Global loss of seafood species.

increase the acidity of sea water by as much as 0.5 pH units by the end of this century, from 8.1 at present to around 7.7. As temperatures rise and the pH decreases, corals will continue to bleach. Sea level rise can also have an effect if it rises enough that the dinoflagellates do not receive the amount of light they need.

5.7.2   Rising Sea Level

Sea level has been rising at a rate of about 10 to 12 cm per century for several centuries at least and probably for several thousand years. The rate of rise has increased during the last hundred years and has accelerated even more during the last couple of decades. Sea level is rising as a result of two phenomena. First, water expands when it is heated. Therefore, if the ocean heats up, sea level will rise. Second, when glaciers melt, the water runs into the sea, causing the level to rise. Apparently, both these things now are occurring. Much of the increase due to melting glaciers is probably coming from midlatitude mountain glaciers as well as high-latitude continental glaciers according to the most recent report by the IPCC.[1] Some prominent examples were shown in the figures in Chapter 3. As for the ice sheets in Greenland and Antarctica, it seems likely that the mass of the Greenland ice sheet will decrease as the Earth warms, but the fate of the Antarctic ice sheet is less certain in the short run. When the climate is warmer, high latitudes warm more than equatorial regions. Part of the reason for this is undoubtedly the increased poleward transport of heat in the atmosphere. By this, I mean that as the climate warms, there will be more moisture in the atmosphere, and more of that moisture coming from evaporation at equatorial and midlatitudes will reach high latitudes and be precipitated out. If it is cold enough, the precipitation will fall as snow, building up high latitude ice sheets. A doubling of CO_2 will probably result in some melting of the Greenland ice sheet, but it may not be warm enough in Antarctica. Don't bet the farm on it, however. The combined effects of the two great ice caps melting, the ocean expansion, and the melting of mountain glaciers will almost surely lead to a rise in sea level. Some scientists have long felt that the western part of the Antarctic ice cap could collapse rather suddenly.[14] The reason for this is that the western part of the Antarctic ice sheet rests not on a land continent, as do the eastern part and Greenland, but on the ocean bottom. This makes it marginally stable, and a warming of the ocean and atmosphere past a certain point may make it disintegrate rapidly. Remember how suddenly the Larson B ice shelf collapsed. The volume of ice that makes up the western Antarctic ice sheet stands at 3.5 million km^3. If it were to melt, the ocean would rise by nearly 5 m, about 17 ft.

Currently, melting of the Greenland ice sheet is thought to contribute about 28% of past sea level, and Antarctica is thought to contribute about 12%. If Greenland should melt entirely, sea level would rise by about 7 m (23 ft). If West Antarctica melts, sea level would rise by 5 m (17 ft). If the rest of Antarctica should melt, sea level would rise about 77 m (253 ft). Many scientists (including me) believe the IPCC has seriously underestimated the danger of sea level rise. I believe that none

GLOBAL MEAN SEA LEVEL

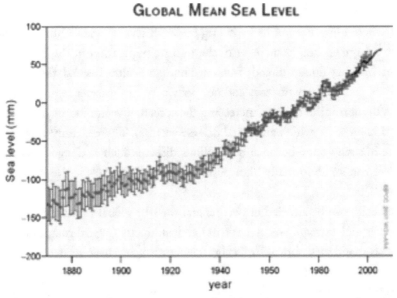

FIGURE 5.8: Annual averages of global average sea level based on reconstructed sea level fields since 1870 (red), tide measurements since 1950 (blue), and satellite altimetry since 1992 (black). Error bars represent 90% confidence intervals.

of us anticipate that the catastrophic cases will occur, but rises of a few feet in the next 50 to 100 years do not seem improbable. A sea level rise of even 1 ft would be devastating to many low-lying coastal areas and islands (Figure 5.8).

For example, about 400,000 people live on the tiny island of Malta. It is highly vulnerable to climate change. Rising sea levels in the Mediterranean during the next century caused by global warming threaten to submerge parts of the island. Also threatened is the island's source of fresh water, which depends on a reservoir of fresh water that lies under the island. Because the fresh water is less dense than salt water, this reserve effectively floats on the sea. As sea level rises, the level of separation between the fresh and salt water will rise, threatening the fresh water supply.

In the Sundarbans, the world's largest mangrove forest, 6000 people have had to be relocated because their land is underwater. The Sundarbans straddle India and Bangladesh at the delta in the Bay of Bengal. Because of the rapid population growth in the Sundarbans islands, if sea level rises even by the modest amount predicted in the report of the IPCC, by the end of the century, millions could be displaced.

If sea level rises by even 1 ft, many large coastal communities could be adversely affected. Southern Florida and the southern Gulf Coast as well as many other coastal areas in the United States and elsewhere in the world would suffer terribly.

5.8   HUMAN HEALTH

New Orleans, where I live, has not had a killing freeze in several years. Quite possibly, as a result, there has been a large and rapid increase in the termite population in the city. Termites cause a lot of problems, but they do not directly influence human health. Instead, the point to be made is that other insects that do carry diseases and normally thrive in warmer regions will likely become more populous in subtropical regions, increasing the spread of vector-borne diseases (diseases that are spread by these insects, which are called disease vectors). There already appears to be a widespread increase in such vector-borne and infectious diseases such as dengue, malaria, hantavirus, and cholera. We had an abnormally high population of mosquitoes in New Orleans during the summer of 1999.

Human health can be affected in two general ways by global warming: directly (e.g., deaths due to heat waves and extreme weather events) and indirectly (e.g., through increases in vector-borne diseases, as mentioned above). It has been estimated that heat-related deaths in large cities could increase by several thousands per year given current predictions of global temperature increase under doubled CO_2. Remember the 1995 summer heat wave in Chicago and the one a few years later in France.

The idea that global warming will have adverse effects on human health has emerged only fairly recently, and many of the potential effects are relatively unfamiliar to public health professionals. Yet it is well-known that the transmission of many infectious diseases is affected by climatic factors such as temperature, humidity, surface water availability, and vegetation. Most of the research concerning how weather and climate will affect the spread of vector-borne diseases is quite recent.

The vector that spreads malaria is the mosquito. Its incidence is affected by temperature, humidity, and surface water. The type of mosquito that transmits malaria does not usually survive where the average winter temperature is below 16–18°C. The incubation phase of the parasite within the mosquito that carries the disease accelerates when the temperature increases even slightly. Malaria is a huge global public health problem, currently amounting to 350,000 new cases each year. Most scientists believe that people in temperate areas would not be seriously affected because public health resources in these regions are sufficient to contain and treat the disease. Developing countries such as Pakistan and African nations would be most affected.

It should be noted, however, that higher temperatures accelerate the life cycles of parasites, and this could result in insects developing resistance to control methods more quickly, and diseases becoming resistant to drugs more quickly.

African trypanosomiasis (sleeping sickness), American trypanosomiasis, onchocerciasis (river blindness), and a number of other serious vector-borne diseases that currently affect people in developing countries (Latin America and Africa for the most part) are spread by mosquitoes and flies. At higher temperature and humidity, the incubation period of the parasite is shorter (so the vector host can infect those it bites more quickly), the lifetime of the vector is shorter, the population density is larger, and the vector feeds more often. All of these effects increase the frequency of infections. Other vector-borne diseases such as dengue (a severe influenza-like disease) and Venezuelan equine encephalitis could easily spread to the southern United States. In 1998, there was an outbreak of St. Louis encephalitis in southern Louisiana. St. Louis encephalitis, a disease carried by mosquitoes, infected 19 people that summer.

The pulmonary disease hantavirus is transmitted by airborne particles of rodent feces. The disease, which was rare in the United States until recently, has been appearing in the southwestern states. Nearly 45% of cases are fatal. Conventional wisdom is that the outbreak follows an unusually strong El Niño. There was more rainfall in the southwest during 1997–1998, and the winter was milder, creating more groundcover and a greater food supply for deer mice, which carry the hantavirus.[15]

Lethal diarrheal disease associated with floods and droughts is expected to rise in east, south, and southeast Asia and rises in coastal water temperature could exacerbate cholera in South Asia.

Of course, if climate change leads to stress on agricultural systems, food shortages could lead to health problems because of malnutrition. Once again, it is the developing countries that will probably be most affected. I have already mentioned heat-related deaths. Heat waves and the resulting hot nights, together with the lack of adequate air conditioning and water as well as marginal health to begin with, make the poor most vulnerable.

5.9   POLAR BEARS

The early breakup of Arctic sea ice is depriving polar bears of weeks of their normal hunting season. The bears use the ice as a platform for hunting seals to build their energy reserves and accumulate fat so that they will remain on land, fasting for a longer time. Data collected in the western Hudson Bay area show the population has declined since 1989 from 1200 to 950, a 22% drop. Historic and predicted future sea ice area changes are shown in Figure 5.9.

5.10   PUFFINS

The biggest colony of puffins in Britain has suffered the worst breeding year on record. The St Kilda archipelago is home to 136 000 pairs of puffins, the largest breeding colony in Britain. The current breeding year is the worst on record in one of the most important seabird breeding areas in northwest region of Europe. The problems that have caused widespread breeding failures among seabirds on the east coast of Britain in recent years have now moved to affect the Atlantic seaboard. The collapse is blamed on global warming and the disappearance from UK waters of the sand eel, the staple diet of many seabirds and a vital element in the marine food chain. A two-degree increase in ocean temperatures in the last 20 years has forced sand eels, and the plankton they feed on, to move to cooler waters.

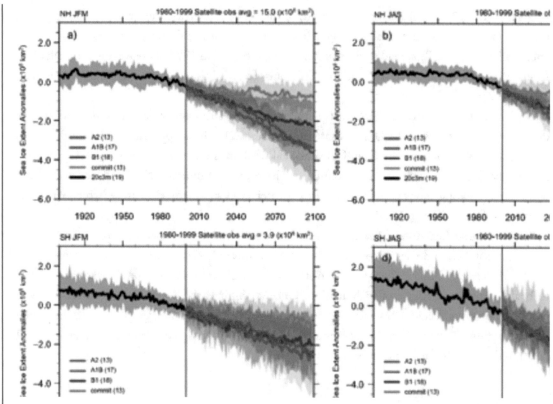

FIGURE 5.9: Multimodel simulations of sea ice cover for January to March and July to September in the northern hemisphere (top) and the southern hemisphere (bottom).

5.11   DIVERSITY OF SPECIES

But it is not just polar bears and puffins that are being adversely affected. Global warming could cause massive species extinctions around the world if it continues unabated. Areas that are particularly vulnerable to climate change include the tropical Andes, the Cape Floristic region of South Africa, southwest Australia, and the Atlantic forests of Brazil, Paraguay, and Argentina because the species in these regions have restricted migration options due to geographical limitations. Slight increases in temperature can force a species to move from its original habitat toward a preferred, usually cooler, climate range. If development and habitat destruction have already altered those habitats, the species often have no safe haven. Climate change could drive more than a quarter of land animals and plants into extinction, according to a major recent study (*Nature*, January 7, 2004).[23] Global warming expected between now and year 2050 will place as many as 37% of all species in several biodiversity-rich regions at risk of extinction.

5.12   ANTARCTIC SPECIES

The Antarctic Peninsula, a key breeding ground for the krill, has warmed by 2.5°C in the last 50 years, nearly five times the global average increase during that time. The resulting decrease in sea ice during the winter has resulted in an 80% decrease in the number of Antarctic krill, a major source of food for animals in the region. The resulting food shortages could threaten Antarctic whales, seals, and penguins. As noted above, krill feed on algae under the ice sheet in the ocean but warmer temperatures during the last 50 years have meant there is less ice and fewer krill. Krill feed on phytoplankton and algae and are, in turn, eaten by fish, ssquid, sea birds, whales, some seals, and penguins. Certain penguin species at several sites in the area where krill populations have declined are already being affected. But the important thing is that the decrease in krill will have a devastating effect on the marine food web in the Antarctic Peninsula region.

5.13   MIGRATION

Global warming is likely to uproot hundreds of millions, perhaps even billions of people, forcing them to leave their homes and creating severe global political and economic challenges. The largest numbers of those forced to move by climate change are likely to be in developing countries, especially those that will experience intensifying drought and sea level rise. If sea level rises by up to a meter this century (the very top of the forecast range), as many as 30 million Bangladeshis could become climate refugees.

Of the 6 billion-plus humans that currently inhabit the Earth, nearly a fifth are threatened directly or indirectly by desertification. China, India, Pakistan, Central Asia, Africa, and parts of Argentina, Brazil, and Chile all have areas with low rainfall and high evaporation that account for more than 40% of Earth's cultivated surface. Closer to home, severe droughts and water depletion in the United States have left nearly a third of U.S. land affected by desertification.

At the same time, hundreds of millions of people live in river valleys where irrigation is fed by glacier melt and snowmelt. As the glaciers gradually disappear, farmers in the Indo-Ganges Plain and in China's Yellow River Basin will most likely face severe disruptions in water availability.

Bangladesh is hardly the only low-lying nation facing devastation because of sea level rise, but it does offer one of the most serious challenges because its large population is spread over a large area that is extremely susceptible and because it is a poor nation. Because such a large portion of the Earth's people live near sea level, a significant rise, even by a foot or two, could cause forced migrations of tens or even hundreds of millions of people. Low-lying coastal zones are also vulnerable to storm surges and increased intensity of tropical storms. Hurricanes Katrina and Rita caused the migration of more than a million people from coastal Louisiana and Mississippi.

The Christian development agency Tearfund has estimated that there will be as many as 200 million climate refugees by 2050 and as many as 1 billion by the end of the century if global warming and its impacts continue.

NOTES AND REFERENCES

Figures 5.1–5.6 and 5.8 and 5.9 are from Reference 1. Figure 5.7 is from Reference 13.

1. Forster, P., Ramaswamy, V., Artaxo, P., Berntsen, T., Betts, R., Fahey, D.W., Haywood, J., Lean, J., Lowe, D.C., Myhre, G., Nganga, J., Prinn, R., Raga, G., Schulz, M., and Van Dorland, R., "2007: Changes in atmospheric constituents and in radiative forcing," in *Climate Change* 2007: *The Physical Science Basis. Contribution of Working Group I to the Fourth Assessment Report of the Intergovernmental Panel on Climate Change*, S. Solomon, D. Qin, M. Manning, Z. Chen, M. Marquis, K.B. Averyt, M. Tignor, and H.L. Miller, eds., Cambridge University Press, Cambridge, UK, p. 130, 2007.

2. Crowley, T.J., and North, G.R., *Paleoclimatology*, Oxford University Press, Oxford, U.K., 1991.

3. Nakicenovic, N., and Swart, R., Eds., *Special Report on Emissions Scenarios. A Special Report of Working Group III of the Intergovernmental Panel on Climate Change*, Cambridge University Press, Cambridge, UK, 2000.

4. For a brief discussion, see Houghton, J., "*Global Warming*: The Complete Briefing," Cambridge University Press, University of Cambridge, U.K., 1994.

5. Holm, L.G., Plucknett, D.L, Poncho, J.V., and Herberger, J.P., *The Word's Worst Weeds: Distribution and Biology*, University of Hawaii Press, Honolulu, HA 1977.

6. Parry, M.L., *Climate Change and World Agriculture*, Earthscan Publications, London, U.K. 1980.

7. Blackman, F.F., "Optima and limiting factors," *Annals of Botany*, vol. 19, pp. 281–286.

8. Meier, W., *Pflanzenschultz im Feldbau*, Tierische Schadlinge und Pflanzenkrankheiten, 8 Auflage, Zurich-Reckenholz, Eidgenossiche Forschungsanstalt fur Landwirtschaftlichen Pfanzenbau, 1985.

9. Karl, T.R., and Knight, R.W., "The 1995 Chicago heat wave: how likely is a recurrence?" *Bulletin of the American Meteorological Society*, vol. 78, pp. 1107–1119, 1997.

10. Groisman, P.Y., "Heavy precipitation in a changing climate," in *Elements of Change—1998*, Aspen Global Change Institute, Aspen, CO. 135 pp., 1999.

11. Caldeira, K., and Wicket, M.E., "Anthropogenic carbon and ocean pH," *Nature*, 425, pp. 365–365, 2003. doi:10.1038/425365a

12. Caldeira, K., and Wicket, M.E., "Ocean chemical effects of atmospheric and oceanic release of carbon dioxide," *Journal of Geophysical Research-Oceans*, p. 110, 2005.

13. Worm, B., Barbier, E.B., Beaumont, N., Duffy, J.E., Folke, C., Halpern, B.S., Jackson, J.B.C., Lotze, H.K., Micheli, F., Palumbi, S.R., Sala, E., Selkoe, K., Stachowicz, J.J., and Watson, R., "Impacts of biodiversity loss on ocean ecosystem services," *Science*, vol. 314, pp. 787–790, 2006. doi:10.1126/science.1132294

14. Mercer, J.H., "West Antarctic ice sheet and CO_2 greenhouse effect: a threat of disaster," *Nature*, vol. 271, pp. 321–325, 1978. doi:10.1038/271321a0

15. Bloom, B.R., "The future of public health," *Nature*, vol. 402(Suppl.), pp C63–C64, 1999. doi:10.1038/35011557

APPENDIX 5.1

Details of SRES Scenarios

The Intergovernmental Panel on Climate Change issued a special report on emissions called Special Report on Emissions Scenarios (SRES) that outlined several possibilities for future emission of CO_2. Several are described below.

SRES A1: A future world of very rapid economic growth, low population growth, and rapid introduction of new and more efficient technology. Major underlying themes are economic and cultural convergence and capacity building, with a substantial reduction in regional differences in per capita income. In this world, people pursue personal wealth rather than environmental quality. The A1 scenario family develops into three groups that describe alternative directions of technological change in the energy system. The three A1 groups are distinguished by their technological emphasis: fossil-intensive (A1FI), nonfossil energy sources (A1T), or a balance across all sources (A1B).

1. World economy grows at 3.3% during the 1990–2080 period. The per capita GDP in 2080 amounts to $76,000 in the developed world and $42,000 in the developing world. The average income ratio between currently developed and developing countries was 13.8 in 1990, and this reduces to 1.8 in 2080. Thus, it is an equitable world where current distinctions between poor and rich countries eventually dissolve.
2. Demographic transition to low mortality and fertility; the world sees an end to population growth. The total population for developing countries reaches 7100 million in 2050 and then declines to 6600 million in 2080, whereas the population in the developed world stabilizes at 1250 million in 2050.
3. Environmental quality is achieved through active measures emphasizing "conservation" of nature changes toward active "management" and marketing of natural and environmental services.
4. Final energy intensity decreases at an average annual rate of 1.3%; transport systems evolve to high car ownership, sprawling suburbanization, and dense transport networks, nationally and internationally; large regional differences in future greenhouse gas emission levels.

SRES A2: A very heterogeneous world. The underlying theme is that of strengthening regional cultural identities, with an emphasis on family values and local traditions, high population growth, and less concern for rapid economic development.

1. World economy grows at 2.3% during the 1990–2080 period. The per capita GDP in 2080 amounts to$37,000 in the developed world and $7300 in the developing world. The average income ratio between currently developed and developing countries was 13.8 in 1990, and this reduces to 5.1 in 2080. Thus, it is a world where income disparities have been reduced by about two thirds of current levels.

2. Rapid population growth continues, reaching a world total of 11 billion in 2050 and almost 14 billion in 2080. The total population of developing countries reaches 9.4 billion in 2050 and 11.7 billion in 2080, whereas the population in the developed world reaches 1.4 billion in 2050 and 1.6 billion in 2080. Social and political structures diversify, with some regions moving toward stronger welfare systems.

3. Environmental concerns are relatively weak, although some attention is paid to bringing local pollution under control and maintaining local environmental amenities.

4. The fuel mix in different regions is determined primarily by resource availability; technological change is rapid in some regions and slow in others; there is high energy and carbon intensity and, correspondingly, high greenhouse gas emissions.

SRES B1: A convergent world with rapid change in economic structures, "dematerialization," and introduction of clean technologies. The emphasis is on global solutions to environmental and social sustainability, including concerted efforts for rapid technology development, dematerialization of the economy, and improving equity.

1. World economy grows at 2.9% during the 1990–2080 period. The per capita GDP in 2080 amounts to $55,000 in the developed world and $29,000 in the developing world. The average income ratio between currently developed and developing countries was 13.8 in 1990 and this reduces to 2.0 in 2080. Thus it is an equitable world where current distinctions between poor and rich countries eventually dissolve.

2. Demographic transition to low mortality and fertility; the world sees an end to population growth. The total population for developing countries reaches 7100 million in 2050 and than declines to 6600 million in 2080, whereas the population in the developed world stabilizes at 1250 million in 2050. High level of environmental and social consciousness combines with a globally coherent approach to sustainable development.

3. Environmental consciousness and institutional effectiveness; environmental quality is high; increasing resource efficiency; reduction of material wastage, maximizing reuse and recycling.

4. Smooth transition to alternative energy systems as conventional oil resources decline; high levels of material and energy saving as well as reductions in pollution; transboundary air pollution is basically eliminated in the long term; low greenhouse gas emissions.

SRES B2: A world in which the emphasis is on local solutions to economic, social, and environmental sustainability. It is again a heterogeneous world with less rapid and more diverse technological change, but a strong emphasis on community initiative and social innovation to find local, rather than global solutions.

1. World economy grows at 2.7% during the 1990–2080 period. The per capita GDP in 2080 amounts to $47,000 in the developed world and $18,000 in the developing world. The average income ratio between currently developed and developing countries was 13.8 in 1990 and this reduces to 2.6 in 2080. Thus, it is a world where income disparities have been reduced by more than four fifths of current levels.

2. Population growth continues, reaching a world total of 9.3 billion in 2050 and almost 10.1 billion in 2080. The total population for developing countries reaches 7.9 billion in 2050 and 8.7 billion in 2080, whereas the population in the developed world reaches 1.2 billion in 2050 and then declines to 1.1 billion in 2080. Social and political structures diversify, with some regions moving toward stronger welfare systems; increased concern for environmental and social sustainability.

3. Environmental protection is an international priority; improved management of some transboundary environmental problems.

4. Energy systems differ from region to region, depending on the availability of natural resources; energy intensity of GDP declines by about 1% per year; technical change across regions is uneven; low level of car dependence and less urban sprawl; transition away from fossil resources; development of less carbon-intensive technology in some regions.

• • • •

CHAPTER 6

The Bottom Line

To investigate the very long-term consequences of increasing CO_2 emissions, the Intergovernmental Panel on Climate Changes[1] has reported the results of an internationally coordinated climate change experiment performed by 23 climate models from a variety of countries around the world. The models were driven according to three of the SRES scenarios shown in Figure 6.1. Also shown in the figure are the 20th century temperature increase and model results with atmospheric CO_2 concentration held constant at year 2000 levels (meaning that CO_2 emissions ceased completely at that time). The lines show the multi-model means, while the shaded areas indicate the ± 1 standard deviation range for the different model means. Not all models were run for each scenario. The colored numbers indicate the number of models run for each scenario in each century. For example,

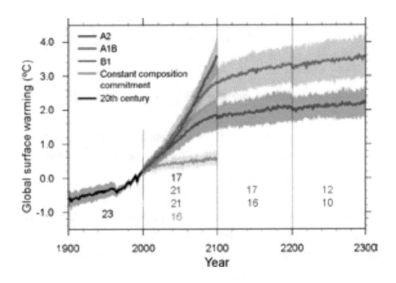

FIGURE 6.1: Multimodel means of globally averaged surface warming relative to 1980–1999 average for scenarios A2, A1B, and B1. Values beyond 2100 are for stabilized emissions. Shading denoted the ±1 standard deviation range.

the blue line represents model results for the B1 SRES scenario. During the period 2000–2100 there were 21 model runs, for the period 2100–2200 there were 16 model runs, and for the period 2200–2300 there were 10 models runs. After 2100 all model runs were run with concentrations held constant at year 2100 values.

Notice that in all the models the temperature continues to increase even after CO_2 emissions cease completely. For example, even if we had stopped emitting CO_2 in 2000, the global temperature would increase by about 0.6°C relative to the 1990–2000 average. In all of the SRES scenarios the temperature is predicted to be considerably higher by 2100, and to continue to rise even more thereafter. A temperature rise of even 2°C will lead to sea level rise of at least 0.5 meters by thermal expansion of the sea alone. Adding the effect of melting glaciers and ice sheets could lead to a devastating sea level rise.

I have already stated that I believe all of the SRES scenarios to be optimistic, given current plans by most governments and views of most individuals and industry leaders. Something like these scenarios, or worse, is inevitable unless governments, industries, and individuals take drastic action to limit greenhouse gas emissions very soon.

And so I end with a brief summary of what we know, what we believe is probable, what we can only consider plausible, and some thoughts about things about which informed and honest people can, and perhaps should, disagree.

It is true that climatologists cannot yet predict the detailed climatic consequences of increased levels of atmospheric greenhouse gases with certainty. There are things, however, that we do know as certainly as one can know anything in science. There are other things that we can predict with very high probability. I now list them.

Conclusion 1. *The atmospheric concentrations of several greenhouse gases are increasing. Included among these is CO_2.*

Conclusion 2. *The increase in atmospheric CO_2 is, for the most part, because of the burning of fossil fuels.*

Conclusion 3. *An increase in the atmospheric loading of CO_2 (or any other greenhouse gas) will result in an increase in the surface temperature of the Earth.*

No scientist who has seriously studied the problem doubts these things. They are simply irrefutably predicted by measurements and the laws of physics.

The central questions that are less well understood are "How much will the global temperature increase?" "How soon?" "What will be the regional distribution of climate change?" Perhaps, the most important question of all, "Does it really matter?"

It is highly unlikely that a doubling of CO_2 in the atmosphere will lead to a global temperature increase of less than 1.5°C. Without this minimal climate sensitivity, it would be essentially impossible to explain the climatic variations that we know have occurred in the past (i.e., ice ages and the like). The global temperature has, in fact, increased by about 0.7 to 0.8°C in the last century. One of the major points made by greenhouse skeptics is that, during this time, there have been periods when the temperature was not increasing. In fact, the temperature leveled off and may have actually dropped a bit during the period between about 1940 and 1970. Temporary decreases in the globally averaged surface temperature of the Earth do not surprise climatologists. There is an inherent "noisiness" in the climate system that need not be caused by any external forcing, such as solar constant variations or greenhouse gases. The climate varies somewhat from year to year and decade to decade and even from century to century naturally. Climate models do not predict a monotonic rise in temperature as greenhouse gases increase. In fact, because the increase in atmospheric CO_2 has been gradual rather than sudden, there is a considerable lag in temperature increase due to the heat capacity of the ocean. It takes a great deal more heat to warm water than to warm air. Current transient climate model runs imply that the climate sensitivity is greater than 1.5°C per CO_2 doubling. It is within the realm of possibility that it is much higher.

Conclusion 4. A doubling of the greenhouse gas concentration in the atmosphere would eventually lead to an increase in the globally averaged temperature of the Earth of at least 1.5°C.
Conclusion 5. The time lag of the climate response is so long that the climate is now only beginning to respond. The 0.7°C increase in global temperature in this century is consistent with higher sensitivity, more likely to be 2.5°C or even higher.
Conclusion 6. High latitudes will warm much more than equatorial regions.

The next important question, "What will be the regional distribution of climate change?," is a central question because it goes to the heart of the "impacts" issues referred to in the fourth question, "Does it really matter?" Physical reasoning and historical evidence as well as all climate models imply that precipitation and soil moisture content distributions will change dramatically, but climate models differ on the particulars. It does appear that some of the agriculturally productive regions of the United States will suffer reduced soil moisture during the summer. High latitude winter precipitation will likely increase, and if this happens, it will likely lead to greater flooding in the spring in some areas, whereas river flow, for example, in the Mississippi River, may be low in the late summer and fall, leading to problems for port cities such as New Orleans. We get our drinking water from the Mississippi River. If the river is too low, salty water can creep upriver and play havoc with our water purification.

Then there is the sea level question. Will sea level rise? We can say with certainty that it will. If the rise is modest and slow, we may be able to adapt. If it is even in the midrange of predictions (about half a meter by 2100), it will be a very expensive and difficult proposition. If the West Antarctic ice sheet collapses or if the Greenland ice sheet begins to deteriorate as rapidly as some now believe, it will be catastrophic.

Conclusion 7. *Climate change will in all likelihood have a serious negative impact on the major food producing regions of the United States. Most climate models suggest that these regions will suffer reduced soil moisture during the spring and summer. It is likely that tropical diseases will spread to higher latitudes. Poor and developing nations will suffer most, but many developed nations will also suffer.*

Conclusion 8. *Coastal areas will eventually be adversely affected by sea level rise, and it now seems quite possible that catastrophic sea level increases may well occur within this century. Low-lying island nations will likely be affected soon.*

Does it matter? Suppose all this happens. It does matter. Any substantial change in the distribution of climate will be disruptive to individual farmers, for example. It is easy to say that regions that can be farmed will simply move northward, but that does not help individuals. Whole nations will suffer food shortages that make present day shortages look like a garden picnic. Major displacements of populations have an enormous impact on a nation's social structure. A fairly rapid change in sea level would have a devastating impact on essentially all nations, both socially and economically. Some have responded that humans have shown an amazing ability to adapt. Irrigation has turned parts of California from dry land to perhaps the most important agricultural region in the nation and the world. Unfortunately, the Colorado River now barely makes it to the ocean. We are pushing the limits of adaptability in many ways.

But what are our choices in a real world? Shall we have a "panel of governments," as suggested by Ross Gelbspan, to reallocate resources and "decide which kinds of renewable, climate-friendly energy are appropriate for different uses and different settings"?

The parties at Kyoto have agreed that certain countries should reduce their greenhouse gas emissions to below 1990 levels by 2010 without giving the slightest hint as to how this can be done. The developing countries, which will very soon be the largest emitters by far, are so far exempted. China is rapidly developing industrially based mainly on coal. Once this infrastructure is in place, they will not be anxious to replace it. Shall we sell them nuclear reactors?

Most of both climate scientists and governments now understand that global warming is real. Representatives of some, but by no means all, large energy companies have decided not to "give up the high ground" and to insist that the science of global warming is insufficiently strong to warrant any change in policy. The U.S. government is beginning to fund programs that are concentrating on impact studies with the intent of showing that the impacts are going to be severe. Some of the work looks more like indoctrination than science. Popular science writers stress either that the science is weak or that global warming will be so bad that we must do something (what?) immediately or face certain disaster.

Almost no one is asking the right question, in my opinion. We need to get past the climate science question. The science is strong. Global warming will occur, and it will be serious, especially if we do not learn how to use energy more efficiently and get away from the use of fossil fuels. The questions that we ought to be debating are social, political, and engineering questions. Continuing the debate about whether global warming is real is not useful. I believe that the real concern of the skeptics has more to do with politics than with science. Massive intrusion by the government into the lives of people has always been anathema for many Americans, and I was once among those who, even living in coastal Louisiana, might prefer to take my chances with global warming rather than adopt the plan of a "panel of governments" as suggested by Gelbspan.

The dirty little secret is that regardless of what we do now, even if we were to stop emitting CO_2 into the atmosphere immediately, we are in for trouble. Just because we have already increased the CO_2 content of the atmosphere, and because it stays in the atmosphere for centuries, the planet will continue to warm for decades, as is demonstrated in Figure 6.1.

Unfortunately, we have gone well past the "tipping point," defined by me as the point after which no matter what we do very bad things will happen because of global warming. The old Chinese curse "May you live in interesting times" is relevant here. Our grandchildren will indeed live in interesting times. This, of course, does not mean that we should not begin as quickly as possible to avert as much global warming as we can.

There are ways to use energy more efficiently, and we should certainly be exploiting them. We also need to explore new ways of generating energy that are relatively clean and environmentally friendly.

These are engineering problems. The engineering community needs to give this top priority. Marty Hoffert has suggested that we treat it as a kind of Manhattan Project and commit to a global thrust to create an energy system that both supplies needed energy to all the globe's inhabitants and does so in an environmentally responsible manner. Some ideas have been put forward in my edited book *Innovative Energy Strategies for CO_2 Stabilization* published by Cambridge University Press in 2002. We need more ideas and a commitment to put them into action soon.

NOTES AND REFERENCES

1. Solomon, S., Qin, D., Manning, M. Alley, R.B., Berntsen, T., Bindoff, N.L., Chen, Z., Chidthaisong, A., Gregory, J.M., Hegerl, G.C., Heimann, M., Hewitson, B., Hoskins, B.J., Joos, F., Jouzel, J., Kattsov, V., Lohmann, U., Matsuno, T., Molina, M., Nicholls, N., Overpeck, J., Raga, G., Ramaswamy, V., Ren, J., Rusticucci, M., Somerville, R., Stocker, T.F., Whetton, P., Wood, R.A., and Wratt, D., "2007: Technical Summary," in *Climate Change 2007: The Physical Science Basis. Contribution of Working Group I to the Fourth Assessment Report of the Intergovernmental Panel on Climate Change*, Solomon, S., D. Qin, M. Manning, Z. Chen, M. Marquis, K.B. Averyt, M. Tignor, and H.L. Miller, eds., Cambridge University Press, Cambridge, UK.

. . . .

Author Biography

Robert G. Watts is the Cornelia and Arthur Jung Professor Emeritus of Mechanical Engineering at Tulane University. He received his bachelor of science degree in mechanical engineering from Tulane (1959), a master of science degree in nuclear engineering from the Massachusetts Institute of Technology (1960), and a doctor of philosophy degree in mechanical engineering from Purdue University (1965). He spent a year at Harvard University studying atmospheric and ocean science. He has worked at the Institute for Energy Analysis in Oak Ridge, TN, and at the International Institute for Applied Systems Analysis in Laxenburg, Austria. His research has concentrated on energy systems and climatology for more than 40 years. He has edited two books on energy and global warming, *Innovative Energy Strategies for CO₂ Stabilization* (Cambridge University Press) and *Engineering Response to Global Climate Change* (Lewis Publishers), and is the author of *Keep Your Eye on the Ball: The Science and Folklore of Baseball* (Freeman Publishing Company) and *Essentials of Applied Mathematics for Scientists and Engineers* (Morgan and Claypool Publishers).

Printed in the United States
by Baker & Taylor Publisher Services